Plant Propagation

ABOUT THIS BOOK

THIS BOOK is intended for everyone who wants to multiply or reproduce the plants they consider desirable. Furnishing a garden can be expensive if every plant has to be bought, and often a great deal of pleasure is lost in buying the finished product.

Though the contents of this book clearly demonstrate the main ways of multiplying plants, emphasis has been placed on the underlying principles which influence plant propagation. It is hoped that readers will, as a result, not just know what to do but will also understand what they are doing. In addition, if a gardener understands the basic principles of plant propagation he will have no difficulty in applying his skills anywhere in the world.

Raising plants by means of seeds, cuttings and other simple vegetative methods are described and illustrated by line drawings, and colour and black-and-white photographs. The influence of environmental factors on all types of propagation is discussed in some detail. A chapter is included on weed and disease control and useful formulae and practical recommendations are included in the appendix. Extensive lists of the most important horticultural genera are provided, together with an indication of the ways in which these are normally propagated commercially.

Half flower of apricot, showing the sexual reproductive organs.

Plant Propagation

K. R. W. HAMMETT

BSc(Hons), PhD

Line drawings by Christine M. Hammett

Photographs by Alan Underhill and the Author

DRAKE PUBLISHERS INC   NEW YORK

ISBN 0-87749-444-4

Published in 1973 by
Drake Publishers Inc
381 Park Avenue South
New York, N.Y. 10016

Library of Congress Cataloging in Publication Data

Hammett, K.R.W.
Plant Propagation

1. Plant Propagation. I. Title
SB119.H285   631.5'3   73-3117
ISBN 0-87749-444-4

Printed in Japan

CONTENTS

LIST OF FIGURES

COLOUR PLATES

PHOTOGRAPHS

PREFACE

THIS IS NOT really a "how-to"-propagate book. I have tried, instead, to explain the basic principles that influence plant propagation, so that anyone who wants to raise his own plants can approach the job with some understanding of the simple practical operations involved.

I believe this approach is important as no two plant species behave in exactly the same way, which means that if a gardener is going to be successful he must employ both intelligence and basic knowledge. Experience is quickly gained when propagation is approached in this way, and then apparently difficult subjects may be tackled with confidence. Such experience can never be adequately conveyed through books alone, as local environmental conditions often have an important influence on details of procedure.

The content of any book is of course limited by its size, and decisions have to be made concerning what is to be included and what is to be omitted. I have elected to deal in some detail with the two most important forms of propagation, namely seeds and cuttings, rather than try to cover a wider field more superficially. Details of budding and grafting, together with a few more obscure methods of propagation, are omitted.

I appreciate that budding and grafting have commercial importance for certain plants, particularly fruit trees, but the average gardener seldom, if ever, needs to use these methods. The enthusiast and student, for whom this book is intended, as well as the home gardener, can refer to larger and much more expensive books for details of these operations. There are a number of books devoted specifically to grafting and budding, and every garden dictionary contains details of these operations, some of which have more relevance to tree surgery than propagation.

The climate of any specific region determines which plants can be grown in that area. Some areas are lucky in that many plants may be grown outside with little trouble, while other areas have a more limited range of plants available to them. It is therefore impossible to treat each subject in detail. However, I hope that the lists in the appendix will be of value as a guide to the usual way in which the most important plants are propagated. Remember, though, that there are often a number of alternative ways possible; those listed are the ones most commonly used by nurserymen. It should also be realised that only the higher plants, namely gymnosperms and angiosperms (flowering plants) are considered. Ferns, mosses and other lower plants are not included.

I have learned more about propagation by writing this book and I hope my readers will also gain something.

Auckland
New Zealand

K.R.W. HAMMETT

INTRODUCTION

THE TERM "propagation" is used to explain the activity by which an individual organism reproduces, multiplies and spreads itself, both in space and time.

The means by which this operation is carried out are diverse, but the very survival of any species depends largely on the possession of an efficient means of propagation. Indeed countless centuries of evolution have selected those plants and animals which most effectively propagate themselves.

In plants there are two basic types of propagation, sexual and asexual or vegetative. It is of fundamental importance to realise that reproduction involving a sexual process is capable of producing plants which differ from their parents. Vegetative reproduction can, however, only produce many identical individuals, which are in fact all derived ultimately from a single plant. The collective term for all the individuals derived from a single plant is "CLONE". For example all cultivars* of rose such as 'Peace' or 'Super Star' are clones as they are propagated by budding, a form of artificial vegetative reproduction.

Man has taken an interest in manipulating plants for his benefit since antiquity. His skill in doing this has had profound effects both for man and the plants with which he has concerned himself. His ability to propagate and grow crops of specific plants has enabled him to give up his nomadic existence and form civilisations. Conversely his farming activities have had considerable effect upon the evolution of certain plants, often to the extent that modern cultivated plants bear little resemblance to their ancestors or other wild plants. Man has done this in three main ways.

First, he has simply selected plants from wild species, which over long periods of time, with successive selection, have evolved into new types. Examples of such plants are bean, tomato, rice and sweet pea.

Second, man has made hybrids between two or more species of plant, the offspring differing from either parent species. Maize, wheat, tobacco and Fuchsia are examples of this type of development.

Third, another group of plants has arisen as rare monstrosities which would have been unable to survive in a natural environment, but have been propagated and preserved by man as they were useful in some way to him. Cabbage, cauliflower and Brussels sprouts are good examples of this type of plant.

Most cultivated plants will either be lost or will deteriorate to less desirable forms unless they are carefully propagated in such a way that the characteristics which make them useful are preserved. To do this effectively a sound understanding of the basic principles underlying propagation is necessary.

Although these basic principles remain unaltered, new propagating techniques or variations of old ones are being developed all the time. Historically, techniques have had to be developed to be able to grow and propagate newly discovered or developed plants. As each new technique has been discovered man has been able to grow a wider range of plants. It is therefore important to realise the historical significance of practical developments made by man in his efforts to propagate plants more effectively.

---

*Cultivar: a cultivated variety as distinct from a naturally-occurring variety.

FIG 1. LIFE CYCLE OF A FLOWERING PLANT

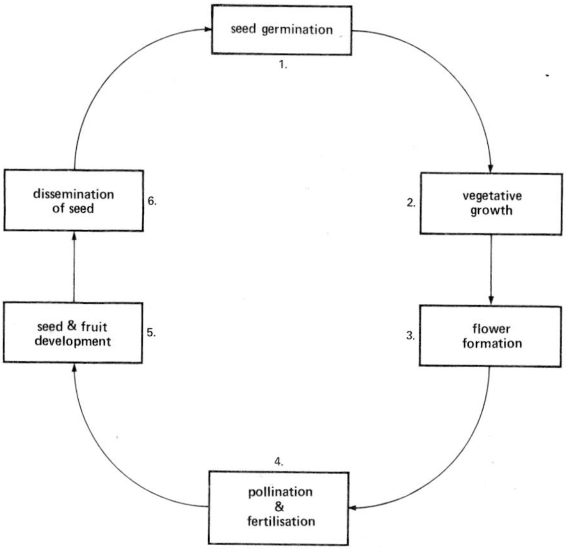

```
                    ┌─────────────────┐
                    │ seed germination│
                    │        1.       │
                    └─────────────────┘
        ┌──────────────┐        ┌──────────────┐
        │ dissemination│ 6.     │  vegetative  │ 2.
        │   of seed    │        │    growth    │
        └──────────────┘        └──────────────┘
        ┌──────────────┐        ┌──────────────┐
        │  seed & fruit│ 5.     │    flower    │ 3.
        │  development │        │   formation  │
        └──────────────┘        └──────────────┘
                              4.
                    ┌──────────────────┐
                    │   pollination    │
                    │        &         │
                    │   fertilisation  │
                    └──────────────────┘
```

FIG 2. PLANT LIFE FORMS

ANNUAL PLANT	HERBACEOUS PERENNIAL	HERBACEOUS PERENNIAL	DECIDUOUS WOODY PERENNIAL	DECIDUOUS WOODY PERENNIAL	EVERGREEN WOODY PERENNIAL
Plant overwinters by seed. e.g. Nemesia.	No aerial growth during winter. New growth from root tubers. e.g. Dahlia.	Plant forms aerial rosette during winter. e.g. Chrysanthemum.	Shrub: more than one stem from ground level. Woody parts added to each year. e.g. Magnolia.	Deciduous tree: one main stem. Leaves are shed in winter. e.g. Elm.	Evergreen tree: one main stem. Leaves retained. e.g. Thuja.

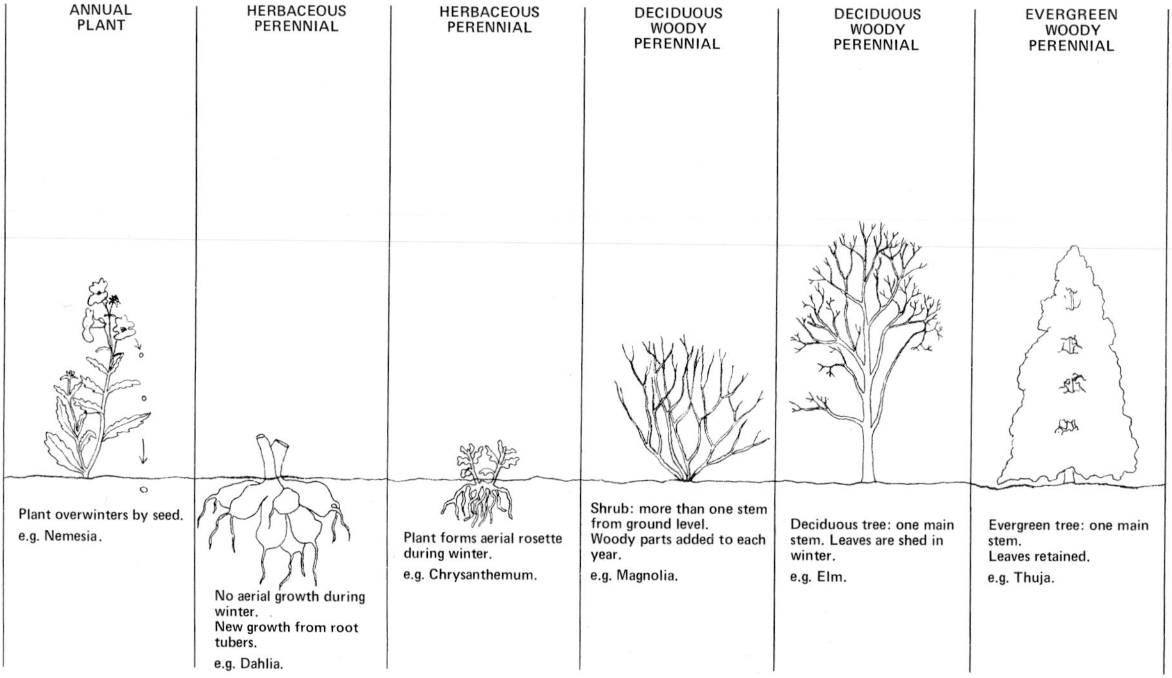

PART I

1

Sexual Propagation

SEXUAL REPRODUCTION involves the union of male and female sex cells, which in the case of flowering plants leads to the formation of seeds.

Life Cycle

Every organism has a life cycle; that of a flowering plant is represented in Figure 1.

The flower is a special structure which enables fertilisation and subsequent seed development to take place. Its parts are formed from considerably modified leaves.

Flowering plants can be classified into several groups, depending on how long it takes them to develop from the seedling stage to flowering. Ephemeral plants can go through their life cycle in a few weeks. Some of the most troublesome weeds are of this type (e.g. *Senecio vulgaris*, groundsel).

Annual plants take a full season to go through this sequence (e.g. *Lathyrus odoratus*, sweet pea). Biennial plants are those with a two-year life cycle. During the first year the plant grows vegetatively, then undergoes a dormant period which usually coincides with winter. In the following year the plant flowers, produces seeds and dies (e.g. *Dianthus barbatus*, sweet william).

Perennial plants live for an indefinite period. Their vegetative phase may extend between one and twenty or more years. Then vegetative growth, flower formation and dormancy occur annually on a cyclic basis. Herbaceous perennials are those in which most of the aerial shoots die during the winter or in dry periods. They survive as tubers, bulbs or similar special structure from which new growth develops in favourable periods (e.g. Dahlia and gladiolus). Woody perennials continue to add to their aerial growth each year, consequently increasing in size. All trees and shrubs are woody perennials. If they lose their leaves when dormant they are said to be deciduous. If they retain their leaves they are referred to as evergreens.

Floral Structure and Function

Flowers come in an incredible range of shapes, sizes and colours, but their structure is basically similar. This basic arrangement is illustrated in Figure 3.

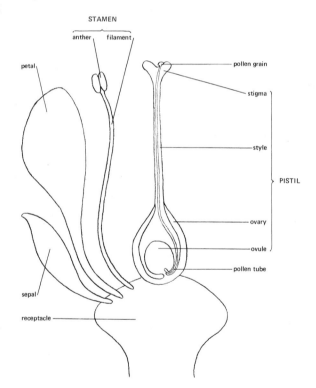

FIG 3. BASIC FLOWER STRUCTURE

The organs making up the pistil are the female part of the flower, and those making up the stamen are the male. The anther produces pollen grains, and the ovary contains an egg cell enclosed within an ovule. During flowering, pollen is transferred from the anther to the stigma. This operation is known as pollination. The actual transference of pollen may occur within a single flower or between different flowers on different plants. After safely reaching a stigma, the pollen grains germinate. Each produces a long tube which grows down the style until it reaches the ovule. Male gametes or sperm cells are produced from the pollen tube and these unite with female gametes within the ovule. This union results in the formation of a zygote and is known as fertilisation.

The zygote develops into a plant embryo while the ovule as a whole develops into a seed. The importance of pollination and fertilisation lies in the fact that male gametes may be derived from one plant and female gametes from another. The union of the two produces an individual which will differ in some way from either of its parents. Suitable combinations of characters enable new plants to be formed which are superior in some specific respect from either parent.

Seed and Fruit Development

While the ovule develops as a seed the ovary develops to become a fruit. Food materials are stored in the seed during maturation. The precise anatomical and physiological details of seed structure and food storage differ as much as floral characteristics. However, in general, to be considered of good quality seeds should be plump and heavy for their size. The initial growth of seedlings depends almost entirely on their food reserves and heavy seeds ensure good germination and vigorous seedlings.

Often fruit and seed are easily separated when ripe—for example, peas and beans from their pods. However, in some plants they are joined in a single unit and the fruit is treated as a seed. Beetroot and wheat are examples where the "seed" is strictly a fruit.

Seed Structure

A seed has been described as a plant packed for transport. It can be considered as consisting of an embryo, food storage tissues and a protective coat.

The embryo is literally a miniature plant consisting of an axis or stem, with growing points at each end; one for the shoot and one for the root. To the axis one or more modified leaves known as cotyledons or seed leaves are attached.

Some plants have only one cotyledon in their seeds, while others have two. So important is this difference to the shape of the resulting plant that flowering plants can be classified into two main groups on this characteristic alone.

Monocotyledons have a single seed leaf and are generally plants with long strap-shaped leaves, such as grasses and lilies.

Dicotyledons have two seed leaves and are broad-leaved plants such as oak trees or beans.

Food may be stored either directly within the cotyledons or in separate storage tissue. Where food is stored in the cotyledons these are usually thick and fleshy.

The seed coat is referred to as the testa. Like the other parts of the seed its structure and origin varies from plant to plant. The function of the seed coat is one of protection. It enables the seed to be handled and transported without injury. It also allows prolonged storage and can have a profound effect on germination.

FIG 4. SEED STRUCTURE (DICOTYLEDONOUS)

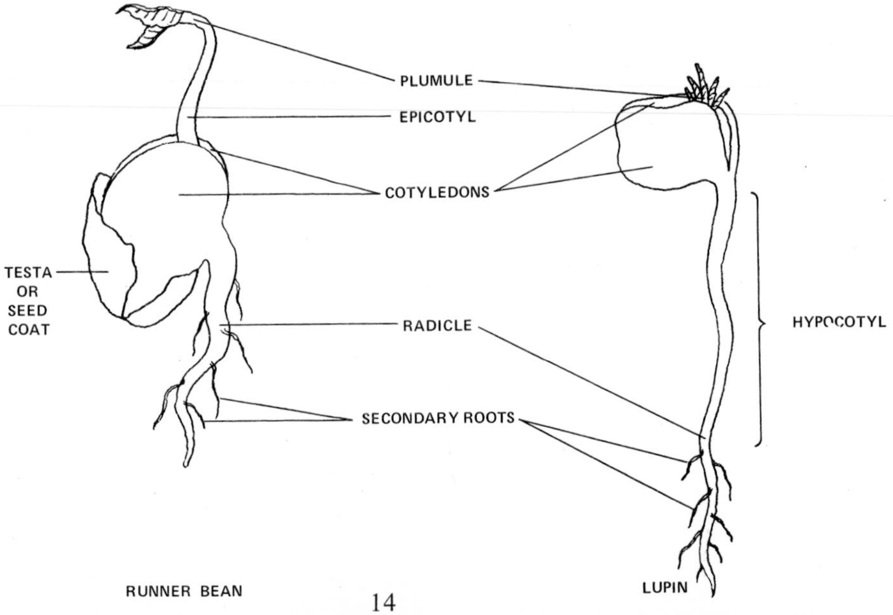

PLUMULE

EPICOTYL

COTYLEDONS

TESTA OR SEED COAT

RADICLE

HYPOCOTYL

SECONDARY ROOTS

RUNNER BEAN

LUPIN

14

2

Seed

IT IS IMPORTANT to realise that more species and cultivars of plants are raised from seed than by any other means of propagation.

Annual and biennial plants have to be grown from seed, which means that almost all farm crops, vegetables and ornamental bedding plants are raised in this way. Many herbaceous and woody perennials are commercially raised from seed even when there are vegetative methods available. It is also interesting to reflect that seedlings are used extensively to provide rootstocks for selected clones of fruit and ornamentals such as rose.

Seed Production

A considerable body of knowledge and experience is involved in the production, cleaning and testing of commercial seed crops. The study of all the methods employed and the reasons behind them is very interesting, but is beyond the scope of this book. Indeed, a good understanding of plant genetics or heredity is needed before many of the operations can be properly appreciated.

However, each year an increasing number of cultivars of seed are offered for sale, which are described as F_1 or First Filial varieties. Such cultivars have had such dramatic effects on the production of some vegetable crops and on the performance of certain ornamental plants that anyone who interests himself in horticulture must try and understand the significance of F_1 seed.

Basically, whenever two distinct cultivars or species are cross-pollinated, the seed and plants resulting from this cross are the first generation after the cross, or the first filial generation. This is indicated by the sign F_1. Subsequent generations are denoted F_2, F_3, and so on.

Geneticists and plant breeders have, however, discovered a special practical significance for the F_1 generation of certain plants and animals. It has been found that when plants which normally exchange pollen between individuals are made to pollinate themselves for a number of consecutive generations they become progressively weaker with each generation of inbreeding. At the same time, if selection for certain specific characteristics is carried out in each generation, after a number of generations the plants will only produce those characteristics and are then referred to as a pure breeding or homozygous line.

When two such pure lines of normally cross-pollinated plants are crossed, the F_1 hybrids are much more vigorous than either of their parents and they are usually more vigorous than the plants from which the parent lines were originally selected. The main advantage with F_1 hybrids lies, however, in the fact that all the F_1 plants from a cross are very uniform in every respect. This uniformity is very important in many ways. In crops that are harvested mechanically it is important that all the plants mature at once. Where vegetables are graded and packed mechanically, it is advantageous to have them of uniform size and shape.

Similarly with ornamental plants it is important to be able to predict their final size and shape in any landscape planning and for them to be uniform in their characteristics and performance. In addition, new combinations of colour and shape are sometimes possible with F_1 hybrids, and they may be earlier flowering.

Often the inbreeding process and the cross-pollination have to be carried out by intricate hand work, which makes F_1 varieties more expensive than traditional cultivars. In addition, F_1 varieties are of considerable advantage to the plant breeder, in that he has a built-in patent system. Seed saved from F_1 plants will not produce uniform plants in the F_2 generation, many of which will be inferior to the F_1 plants. Therefore the F_1 varieties have to be produced each year from the pure lines maintained by the breeder and known only to him.

F_1 hybrids have had considerable impact on the following plants: maize, tomatoes, onions, sorghum, cabbage, petunias, African and French marigolds, cucumber and melon. It is also interesting to reflect that the tremendous revolution brought about in the poultry industry of various countries has largely been the result of F_1 hybrids.

Seed Storage

Seeds are usually stored for varying periods after harvest. Viability or the ability to germinate and produce plants depends on various factors. Among these are the initial viability at harvest after handling and cleaning, the rate of loss of viability of the species and the conditions under which seed is stored.

Some species have seed which is naturally very short lived and cannot be stored, and other species have long-lived seed.

Basically the storage conditions which best maintain seed viability are those that slow down the chemical life processes or metabolism without injuring the embryo. This involves finding the best moisture content of each species of seed, something which varies widely, and reducing the storage temperature.

Fluctuations of moisture content and temperature are harmful. Leading seedsmen nowadays have carefully-controlled seed storage rooms, and an increasing amount of seed is sold either sealed in plastic and foil envelopes or in sealed tins. The seed is brought to the correct moisture content before being sealed. Generally, seed bought in such containers is a far better buy than that sold in ordinary paper envelopes.

When buying seed sold only in paper packets it is wise to buy the freshest seed possible. In some countries the seed trade makes a practice of printing on its packets, "Good until such-and-such a date". Usually this date is set two to three years after the date of packing. Such a practice enables old seed to be carried forward to be sold in the next season. The age of seed can be estimated, as the nearer the date on the packet is to the current date, the older the seed.

In other countries the seedsmen withdraw all unsold seed from the shops and replace it with fresh seed. Fewer seedlings can be expected from old seed and these are usually less vigorous than those from fresh seed.

Cheap seed is rarely a good buy, and it is worthwhile taking trouble to obtain the best seed available.

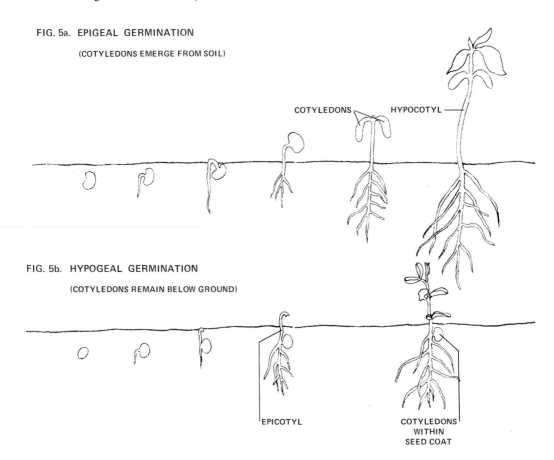

FIG. 5a. EPIGEAL GERMINATION

(COTYLEDONS EMERGE FROM SOIL)

COTYLEDONS   HYPOCOTYL

FIG. 5b. HYPOGEAL GERMINATION

(COTYLEDONS REMAIN BELOW GROUND)

EPICOTYL   COTYLEDONS WITHIN SEED COAT

3

What Makes a Seed Germinate?

The Process of Germination

After a seed is separated from its parent plant and during the time it is stored and transported, it is said to be quiescent, or more commonly, dormant. When the embryo starts to grow and produce a seedling, the seed is said to germinate.

For a seed to germinate the embryo must be viable or alive and able to germinate; any physical or chemical barriers within the seed must have disappeared, and the external or environmental conditions must be favourable.

Seeds of different species have different environmental requirements, but water, warmth and oxygen are fundamental and sometimes light is needed as well. Usually it is interaction between these factors which stimulates germination.

Germination is a complex process which can be divided up into phases. Broadly, these are water uptake, start of chemical activity, mobilisation of food reserves, and growth of seedling. The seedling depends entirely upon food reserves in the seed until its leaves can function and produce its own food.

There are two basic types of seed germination. In one the seed is lifted out of the soil and the cotyledons turn green. This is called epigeal germination. In the other, the seed and cotyledons remain below ground. This is hypogeal germination.

The growing point of the root is the radicle, the growing point of the shoot the plumule. The area of seedling stem above the cotyledons is the epicotyl and that below is the hypocotyl.

Some Reasons Why Seed Will Not Germinate

Many plants produce seeds which will germinate as soon as harvested. In fact a few will, in wet weather, germinate while still on the parent plant. However, some seeds have delaying mechanisms which enforce a period of dormancy.

In nature, germination-controlling mechanisms are important because they spread germination over an appreciable period of time. This means that if a proportion of seed germinates during an unfavourable time and is killed there will still be seed available to germinate at a future and perhaps more favourable time.

Under cultivation, where environmental conditions are to some extent controlled, there is no advantage in delayed germination. Indeed, from the grower's point of view it is a considerable nuisance to have to try to overcome such mechanisms so that he may get a good uniform germination. As a result the development of cultivated plants has involved the selection of cultivars which do not, usually, have complex germination problems.

Most difficulty in germination is found in seeds of trees and shrubs of native plants or those recently introduced into cultivation.

Different mechanisms and combinations of mechanisms are found which control dormancy in different species. When trying to grow seeds that have an imposed dormancy it is important to know which type of mechanism is operative so that it may be overcome.

1 Hard coated sweet pea seeds. The seeds on the left are unchipped and unswollen. The seeds on the right are chipped, swollen and germinating. Both batches were placed on damp blotting paper in a warm room for four days.

The simplest germination delay-mechanism is that of a hard seed coat, which prevents water being taken up. In nature such coats are broken down by agencies such as mechanical abrasion, alternate freezing and thawing, attack by micro-organisms or passage through the digestive tract of birds and animals. When cultivated, quite a number of species which have hard coats need to be broken artificially. For agricultural seeds such as clover the seeds are scratched or scarified mechanically *en masse* by a machine called a scarifier. For large horticultural seed such as hard-coated cultivars of sweet pea, the coat of each seed is cut individually with a knife or file on the side farthest from the hilum or point at which the seed was attached to the ovary. This is known as chipping.

With mixtures of varieties it is a good idea to soak the seed overnight. Those with permeable seed coats will swell and sink to the bottom, and those with hard coats will remain unswollen and may then be chipped.

Occasionally hot water or acid treatment of hard-coated seed is employed, but as both involve an element of risk—to the seed and the operator—they are not unconditionally recommended to the amateur.

A second type of dormancy exists where seeds have specific temperature requirements after harvest, which must be fulfilled before they can germinate. There are many variations and combinations of requirements. A well known requirement occurs in a number of autumn-ripening tree and shrub species with moist seeds. These must be chilled overwinter before they can germinate in the spring. This has led to the horticultural practice of "stratification" where seeds are placed between layers of moist peat or sand and exposed to low temperatures outdoors or kept in a refrigerator just above freezing point for several months. Apple seed is an example of a species with a stratification or "moist-chilling" requirement.

Another type of dormancy is controlled by the presence of chemical inhibitors either in or surrounding seeds. Seeds of this type can be germinated only when this inhibitor is removed or inactivated. This is usually carried out by soaking the seed; the inhibitor is then leached away.

A final group simply requires a period of time in dry storage, during which some after-ripening processes can take place within the seed, enabling it to germinate.

External Requirements

As mentioned, water, oxygen and warmth are the basic requirements for germination of any seed.

WATER

Seeds have a considerable absorbing power for water. In storage they are capable of absorbing it from the air. The availability of water influences both the rate of germination and the percentage of seedlings which become established. Different plants have different requirements regarding the amount of water available in the soil before they can germinate. Some can germinate only when there is a lot of water in the soil and others when there is not too much.

TEMPERATURE

Favourable temperatures are required for seeds to germinate. Plants can be conveniently classified, regarding temperature requirement, as follows: a. Those whose seeds germinate only at relatively low temperatures. b. Those which germinate only at relatively high temperatures. c. Those which germinate over a wide range of temperature.

These requirements are important in nature as they help to determine a plant's adaptation to a specific environment. They also determine at which time of the year seeds will germinate. Plants from cold regions will often only germinate below a certain temperature, while those from the tropics need high temperatures to germinate. Similarly, this mechanism explains why different species of annual weeds tend to appear in large numbers at set times of the year.

The complex germination/temperature reactions of many plants have been studied by scientists, but much of this subject is beyond the scope of this book. It is interesting, however, to note that often an alternating day and night temperature gives better germination than a constant temperature.

OXYGEN

Oxygen is required within the seed for respiration and other complex metabolic processes, many of which are still not fully understood. A few plants, especially water and swamp plants, can germinate when oxygen levels are relatively low. However, the vast majority of plants require a good supply of oxygen before they can germinate.

LIGHT

Some seeds need to be germinated in the dark, a few others need light stimulation before they will germinate. However, light has much more general importance to the seedling. In the darkness of the soil, the hypocotyl or epicotyl is elongated and etiolated (almost colourless) with the leaves unexpanded. There is a hook at the end of the plumule which pushes through the ground. When the seedling shoot emerges into the light, shoot elongation decreases, the plumular hook straightens out, leaves expand and normal aerial growth takes place. The sooner this stage is reached the sooner the plant is independent of the food reserves of the seed.

4

Raising Seedlings

THE OPERATION of raising seedlings involves little more than obtaining good seed which is ready for germination and then providing the external conditions discussed in the last chapter.

Sowing Outdoors

Far more plants are raised by sowing seed outdoors than by any other method of propagation. This is an important point, as it is easy in a book of this type to put too much emphasis on elaborate and expensive structures and techniques.

Two techniques of outdoor sowing can be used. The first is direct sowing into permanent positions and the second involves the use of a seed bed from which seedlings are transplanted.

Direct sowing is the simplest and most economical form of raising plants since it involves no special facilities and no individual handling of seedlings. It is the standard method for field crops and many vegetables. Also, forest regeneration is widely achieved through direct seeding. Many annual ornamental plants are easily raised by this method, especially those which resent being transplanted, such as zinnias.

Often seed of annuals is described by seedsmen as either being a hardy annual or a half hardy annual. Such a distinction only really applies to the area in which it is made and is usually used by British seedsmen. Basically, it means that those seedlings described as hardy may be sown directly into their flowering positions in autumn or early spring before the danger of frost is over. Half hardy annuals, on the other hand, are damaged or killed by frost, but require a longer growing season than is available if sowing is delayed until all danger of frost has passed. Half hardy annuals are therefore raised under the protection of glass and planted out later. In regions that have little frost, there is a sufficiently long growing season to enable many half hardy annuals to be sown directly, with great success.

When direct sowing is used in the flower garden a distinct area should be marked out. This enables any weed seedlings to be more easily recognised and removed. Later when the seedling plants have started to become established, their numbers may be thinned out to give each plant sufficient room for development.

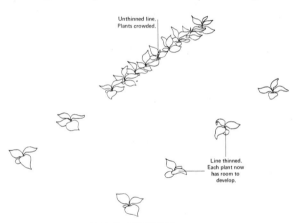

Unthinned line.
Plants crowded.

Line thinned.
Each plant now
has room to
develop.

FIG 6.   THINNING LINES OF OUTDOOR SOWN SEEDLINGS

Direct seeding has the advantage that individuals receive no check in growth associated with transplants and consequently the time between seeding and flowering or harvest is less than by other methods. However, if high priced seed is used it can be wasteful, and if the ground contains a lot of weed seed control can be difficult, especially in the flower garden.

A recent commercial development has been the development of precision direct sowing. Here the very highest grade seed is fastened to soluble plastic tapes at specified intervals. The tapes are laid into shallow drills by a machine similar to a cable layer. Savings in labour, easier weed control and lack of damage to the plants have resulted in greatly increased yields. Small tape packs are being made available to home gardeners in the USA.

When to Sow

Outdoor sowing, whether direct or by using a seed bed, uses rainfall as its main supply of water; sunshine for its energy source; and the condition of the soil to determine the relative availability of water, temperature and oxygen. A seed bed usually only differs from other areas of a garden in that special care is taken in its preparation to specially suit the germination of seed and the development of seedlings.

The time of the year when sowing is made out of doors is obviously a very important consideration and the decision must be made by the grower in relation to his own area. For most plants the spring is usually the most favourable time, as rainfall can be expected to be adequate, temperatures will rise during subsequent months, and the days will become longer. The winter months are generally too cold or wet and the summer months are often too hot or dry. The autumn has similar conditions to spring and can be used, but remember that unlike spring the conditions become progressively worse with time rather than better.

In cooler climates the summer is used for raising biennial plants such as sweet william and wallflowers. In every case the grower has to balance what he wants to grow with his own situation and resources. Notes such as these can only guide a personal decision.

Soil Preparation

Basically, soil into which seeds are to be sown should be well drained, but retain supplies of moisture; be sufficiently open to provide oxygen and good gas exchange with the air. It should also be of such a nature that it does not allow wildly-fluctuating temperatures. Fortunately, measures to provide any one of these requirements help with the others.

If land is badly drained on a farm or garden scale, provision of land drains will obviously have to be made to physically remove excess water. If specific beds or small areas have soil which tends to become waterlogged, something such as coarse sand or an organic material such as compost, peat or farmyard manure should be incorporated. This will have the effect of making the soil more open and thus more easily drained as well as improving oxygen supply. The addition of organic material aids water and nutrient retention. It should, however, be realised that excess soluble salts from artificial fertilisers can be very harmful to seeds and seedlings. Chemical fertilisers should therefore be used sparingly where seed is to be sown direct. It is advisable to apply a light initial dressing several weeks before seed is to be sown and then use side dressings later in the season when the plants are established.

Soil should be fairly firm for seed sowing. If it is

2 Raised frames used for hardening-off bedding plants. They have the big advantage over the low frames (Figure 7), in that they minimise the amount of bending required by the nursery worker. A better watch of the plants may also be made as they are nearer eye-level.

too "fluffy" fluctuations in moisture content and temperature can occur which harm germinating seeds. In addition, such soil provides poor anchorage for mature plants. A firm seed bed is especially important for grass, members of the cabbage family, and onions.

However, all seed should be in close contact with soil particles to ensure the most efficient moisture uptake prior to germination.

Sowing Indoors

Exactly the same principles apply to growing undercover as outside. The difference lies in the amount of control the grower has over environmental conditions. Usually a glasshouse is used for propagating and frames or lathhouses are used for hardening off prior to planting out or selling.

The question of glasshouse design is an extensive subject, which is continually evolving. Many books have been written about glasshouses and some of them should be consulted before a glasshouse is built. Similarly, many systems of heating are available, each of which has points for and against. However, as air exchange and movement are desirable as aids to controlling temperature, humidity and certain foliar diseases, an electric fan heater is probably the simplest and most effective. In this context large side and roof ventilators are today considered desirable to ensure that the air is changed frequently within each glasshouse.

Maximum light is required for established plants in glasshouses, especially during the winter months. However, uncontrolled radiation from the sun can very easily and quickly raise the temperature within a glasshouse to dangerous levels, especially in one of small dimensions. Excessive temperatures are particularly dangerous to seedlings and can result in damage very similar to that produced by damping-off fungi. Shading of both the whole glasshouse and individual seedling containers is often necessary in areas where light levels are high.

Plastic film and sheets of various types are being used increasingly for protected growing, both on a commercial and home garden scale. Plastic film has the advantage of being lower in cost and less heavy than glass, but light transmission is not as good as in clean glass. Plastic film has the disadvantages of being short lived and fairly easily torn. Humidity also tends to build up under plastic, leading to condensation drip. Ideally, wire mesh should be used to sandwich the film, thus adding strength and rigidity.

Translucent fibreglass sheets are now used to some extent.

Cold frames are very simply constructed and a typical one is shown in Figure 7. These are best situated near the glasshouse in a position protected from the wind. As their name suggests, they are not normally provided with any artificial heat. They are used primarily for conditioning or hardening seedlings and rooted cuttings before planting or selling. They are also useful for raising biennial plants at a time of the year when artificial heat is unnecessary.

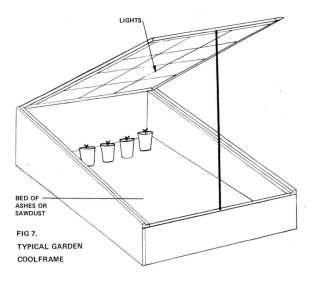

LIGHTS

BED OF ASHES OR SAWDUST

FIG 7.
TYPICAL GARDEN COOLFRAME

3 *Left:* Well-constructed greenhouses built of wood and polythene sheet. The whole of the side walls may be lowered or raised as required.

21

When young plants are first moved from a glasshouse the lights or the top of the frame is kept closed for a few days. When the plants start to become established, the lights are gradually raised.

Lathhouses are commonly used in warm countries in place of cold frames, as well as for general growing and holding of shade-loving plants. Basically, they consist of a framework covered either with thin strips of wood, with a gap of similar width in between, or with one of the synthetic shade cloths. Usually such houses have a headroom of seven to eight feet.

Growing Media

In the same way that temperature, light and humidity are to some extent controlled indoors, growing plants in containers affords the grower the opportunity to provide an ideal rooting medium for his plants. Unlike outdoor growing where the existing soil has to be modified to make it more suitable, the grower can start from scratch.

A growing medium must be sufficiently dense to afford good anchorage, provide ample water without frequent waterings, have good drainage to allow plenty of soil air, be free of weed seeds and disease organisms, and must not have too high a salt or nutrient level.

Traditionally gardeners have made up their own seed and potting composts, each swearing by his own recipe. Indeed many of them have had as many ingredients as a Christmas pudding. However, batches of such concoctions vary considerably each time they are made. This may be acceptable to the amateur, but commercial producers need a degree of standardisation and uniformity, particularly regarding disease control. For this reason a number of research stations have looked at the problem and come up with specific recommendations.

Basically, there are two types of growing medium, those based on soil and those from which soil is omitted. Today the soil-less media are gradually replacing the soil-based ones. This has been a natural progression as materials such as peat, vermiculite, perlite, sawdust, leafmould and sand have had to be used to make soil suitable for container use. By omitting the soil altogether, control of texture, nutrient balance and disease control are considerably simplified and batches more easily standardised.

Materials do, of course, vary quite a lot from country to country, and gardeners probably reflect the diversity of horticulture in wanting to be a little different from the next fellow. For this reason there has been a tendency for each mixture to become elaborated for specific purposes. Many mixture formulae are therefore to be found in gardening books.

Details of basic John Innes and University of California growing media are included in the appendix. Both formulae have proved themselves to be highly successful for a very wide range of subjects in many areas of the world.

Containers

There are a multitude of containers, both designed for the purpose and improvised, which may be used for growing seedlings. All that is required is something to hold the growing medium, enable water to be added and have adequate drainage to prevent waterlogging.

The normal practice is to sow seed into a relatively small container and when the resulting seedlings are large enough to be handled, they are transplanted or pricked out, either singly into small pots or spaced into boxes or punnets. Traditionally, shallow wooden

FIG 8. SOWING SEQUENCE

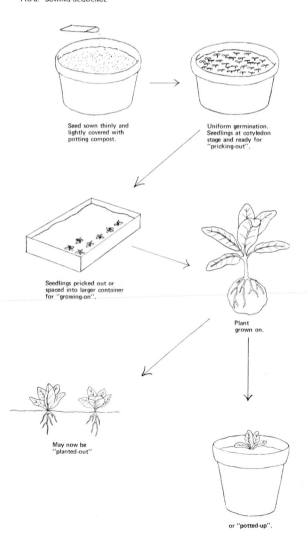

Seed sown thinly and lightly covered with potting compost.

Uniform germination. Seedlings at cotyledon stage and ready for "pricking-out".

Seedlings pricked out or spaced into larger container for "growing-on".

Plant grown on.

May now be "planted-out"

or "potted-up".

seed boxes are used. In the United Kingdom and New Zealand forty-eight to sixty plants are put into each. In the USA, larger trays are often used.

A disadvantage with this method is that the boxes are heavy and where sixty plants are pricked out they can easily become drawn through over-crowding. In addition, when vegetable and bedding plants are sold, many buyers do not require a whole box, which necessitates digging into the plants to sort out smaller numbers. This is a messy business for the shop assistant, the shop floor and the customer. Also, damage is easily caused to plants at this stage and those remaining are less easy to look after. Finally, the customer needs to plant his purchase fairly quickly after buying, as loose plants are invariably wrapped up as a bundle in newspaper.

In a number of countries many alternatives are used successfully and wooden boxes are becoming old fashioned. Among the many interesting alter-natives are multipots, made both from plastic and peat fibre. Here small units, containing either single plants or a small number, are joined rather like postage stamps. A precise number of plants can then be torn off with none of the disadvantages of the wooden box. Another answer to the problem is the punnet. This is a container, rather like a strawberry punnet, with ample drainage holes. Each holds twelve plants, although reputable nurseries will put in fourteen to fifteen to make sure each buyer gets a minimum of a dozen. The big advantage to the gardener is that if he buys one of these containers he need not plant for several weeks if the weather turns unfavourable.

As with seed trays, pots used for growing larger plants have undergone changes in the last ten years or so. Traditionally, pots have been made of clay. For some subjects they are still very good and for decorative work they have the advantage of being a

4 A well-grown punnet of bedding plants. Full cultural details are given on the attached display card.

natural material, and hence aesthetically pleasing. However, they have the disadvantages of being heavy, are porous and lose moisture easily, and are easily broken. After continued use they accumulate salts which can reach toxic levels and which cause green slimy algae and moss to grow on the outsides.

Plastic pots are today much more commonly used and overcome all the disadvantages mentioned for clay pots. However, if he has been using clay pots a grower needs to rethink his whole approach to them. It is no good thinking they behave just like clay pots. Plastic pots are easy to overwater and the grower may need to make his growing medium more open or coarse, and water less frequently. Never-theless, once mastered, plastic pots can produce excellent plants.

The home gardener frequently makes use of plastic drinking cups and icecream tubs. These are very good provided a sufficiently large drainage hole is burned into them with a red-hot poker. On a commercial scale plastic bags and tin-coated cans are being increasingly used for larger subjects. Shrubs are grown in them right from the young plant stage. Such con-tainer-grown plants have considerable advantages over those grown on in the field and then lifted for sale during winter. They may be planted at any time of the year and do not suffer the shock of being wrenched and lifted from a nursery line and trans-ported either bare rooted or with their roots wrapped in hessian.

A distinction should be made between container-grown plants which have been grown in containers

5 Pot of seedlings at the stage when they should be pricked out to give each individual enough room to develop.

all their lives and containerised plants. Containerised plants are usually unsold field-grown plants which are put into containers at the end of the winter selling season in the hope of making a sale during the following flowering season.

Probably the only real disadvantage of container-grown plants is the possibility of tree roots following the shape of the container instead of spreading. This, however, only applies if stock remains unsold and is consequently kept in the container too long. Often a smaller, younger, cheaper plant is a better buy than larger, older ones.

1 Hypogeal germination of *Phaseolus coccineus*. Cotyledons remain below soil level.

2 Epigeal germination of *Phaseolus vulgaris*. Cotyledons are lifted above soil level.

3 Dahlia tuber sprouting. The shoot is rising from the crown of the tuber and is suitable for use as a cutting.

4 *Above* A runner bean flower *(Phaseolus coccineus)*.
5 *Right* Carrot flowers *(Daucus carota)*. Most vegetables are flowering plants, and in fact must flower for seed to be formed. Some vegetables have very beautiful flowers when viewed close up, as is the case with the scarlet-runner bean and the carrot.

6 Modern sweet pea varieties. Sweet peas are annual plants and are raised from seed each year. Sweet pea varieties are true breeding lines.

7 Premature sprouting of seeds in the seedhead of Helichrysum while still on the parent plant.

9 Cuttings lined up and labelled within the propagating frame.

10 *Below:* Seedlings being pricked out in a punnet. Note that the seedlings are being held by the cotyledon leaves. This avoids any possible damage to the seedling stems through rough handling.

11 Bedding plants in a punnet, attractively produced and labelled.

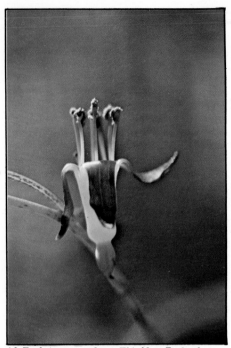

excellent ground cover in areas with no severe frosts. The flowers point directly upwards, unlike most Fuchsias, and have blue pollen. Being a species and not a hybrid it may be propagated from seed, the seedlings being essentially like the parent plant. It may also be very easily propagated vegetatively by cuttings.

12 *Fuchsia procumbens.* This New Zealand native plant is by far the most unusual Fuchsia species and is much better known overseas than in its country of origin. The plant makes an

13 *Above* Modern Dahlia varieties. These varieties are clones and are propagated vegetatively. New varieties are raised from seed, only the best seedlings being maintained as clones.

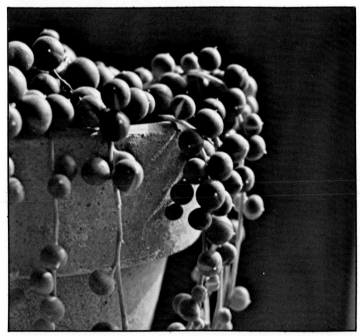

14 *Senecio Rowleyanus.* This unusual succulent plant has almost completely spherical leaves borne in long strings on the trailing stems. It is easily propagated by cutting into short lengths and laying them on the surface of moist soil in pots. Being succulent it is able to form new roots before the pieces dry out.

15 Coloured leaves of the zonal pelargonium (geranium) 'Mrs Pollock'. Attractive leaf characteristics such as these would not be consistently reproduced in young plants raised from seeds taken from a plant like this. Therefore such characteristics are maintained by vegetative propagation of clones.

PART II

5

Vegetative Propagation

VEGETATIVE or asexual propagation is possible because the vegetative parts of many plants have the capacity for regeneration. The ability for vegetative regeneration varies widely between species and the parts of plants. Stem cuttings from many plants are able to form a new or adventitious root system. In some plants roots are able to form new shoots, and in others leaves can form both new shoots and roots. Man has also found it possible in some cases to graft parts of one plant on to the roots of another.

Under carefully-controlled conditions, whole new plants have been formed from single isolated cells, so it seems clear that any living cell of a plant has all the genetic information needed to regenerate a whole plant.

The Importance of Vegetative Propagation

As mentioned in the Introduction, cultivars propagated by vegetative means are clones. Every individual produced is genetically identical to the original plant from which the clone was started.

Many of our most important plants have such a

6 Plantlets developing along the edge of the leaves of a Bryophyllum plant. This is a natural form of vegetative propagation, albeit an unusual one.

7 Plantlets of Bryophyllum establishing themselves at the base of their parent plant after falling from the leaves. The lower leaves of the parent plant have lost all their plantlets.

FIG 9. RUNNER (STRAWBERRY)

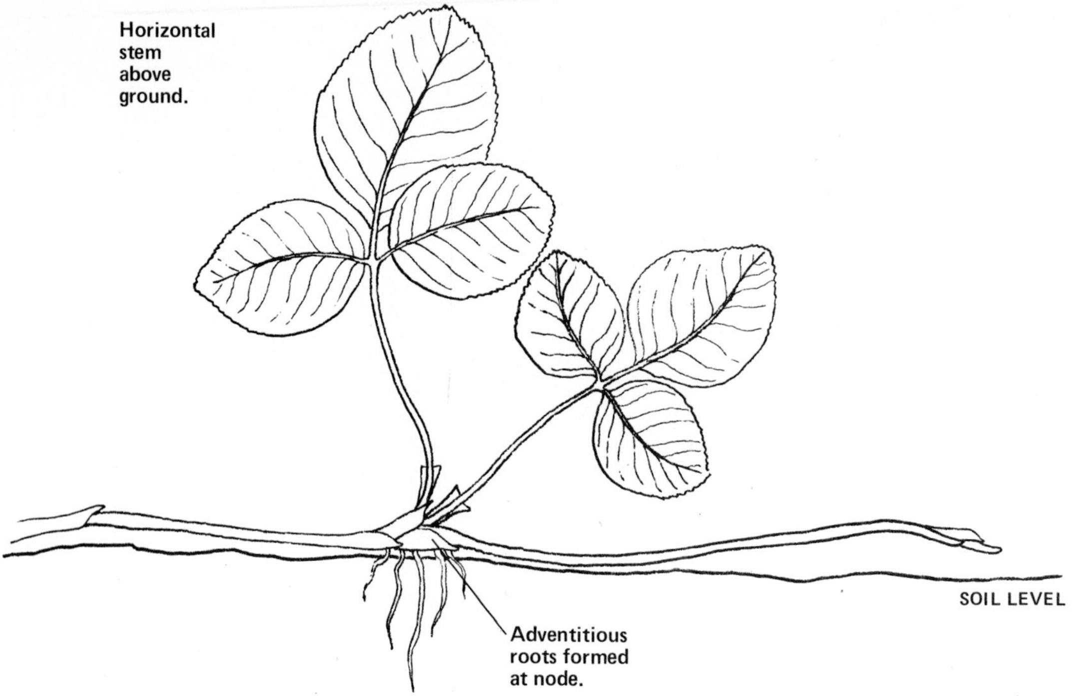

Horizontal
stem
above
ground.

SOIL LEVEL

Adventitious
roots formed
at node.

FIG 10. RHIZOME (MINT)

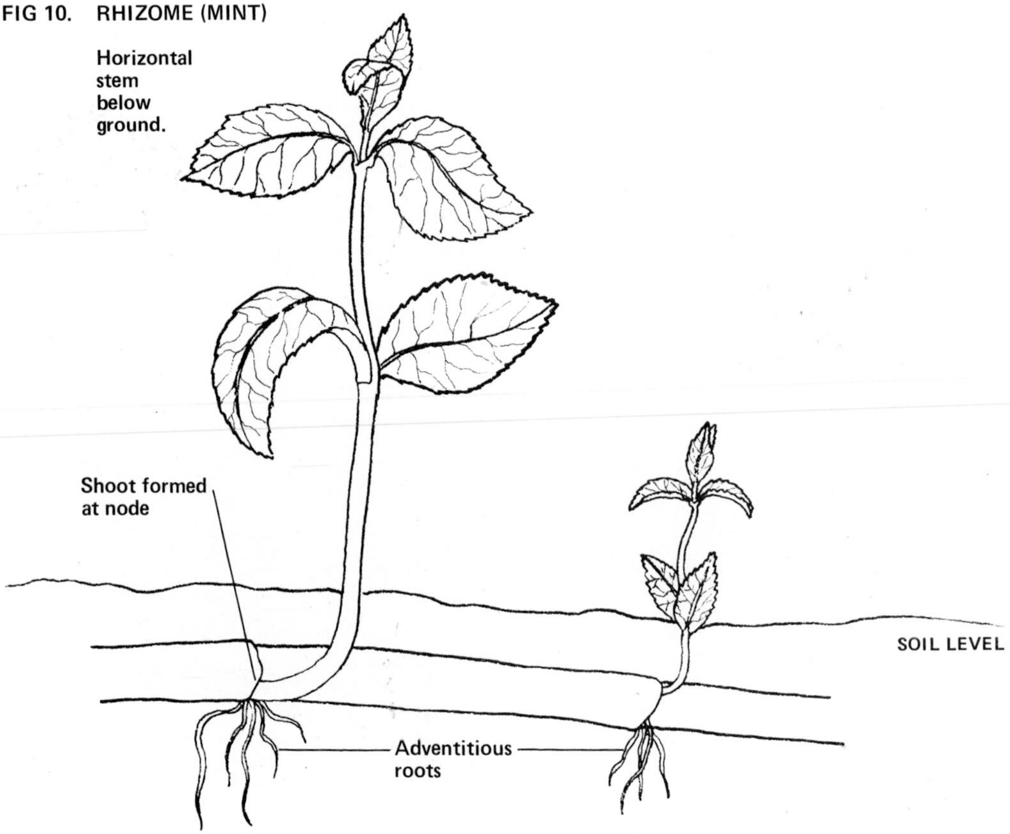

Horizontal
stem
below
ground.

Shoot formed
at node

SOIL LEVEL

Adventitious
roots

complex history of hybridisation and cross-pollination that seedlings raised from them would be extremely variable. Parental characteristics would be combined in endless permutations and few if any of the seedlings would have as many virtues as the parent cultivar. Therefore when a particularly favourable combination of characteristics is found in a seedling it is important to preserve it as a clone. This can be achieved only by vegetative means.

Besides this central reason for using vegetative propagation various other considerations often make it desirable. Some cultivars produce no viable seed (e.g. certain bananas and oranges), and these have to be propagated asexually.

Often it is quicker and easier to take cuttings than to raise seedlings, although of course the converse is true in other cases.

In some plants there is a long juvenile period in which no flowers or fruit are formed. Also, some juvenile forms have undesirable features such as thorniness. These objections can be overcome by propagating from mature plants.

Clones are often maintained for very long periods. Fruit tree cultivars, which have a long life cycle, present a number of examples where clones have been maintained for more than 200 years.

Natural Forms of Vegetative Reproduction

Clones exist in nature where plants have structures such as bulbs, corms, rhizomes, runners or stolons. Though these structures certainly do increase the number of individuals, their primary function is often something different; for example, food storage or as a means of overwintering. Even the plants with the greatest ability to spread vegetatively can affect only a relatively small area by this means alone. Without the intervention of man they are not really effective in disseminating a clone and almost invariably the plant produces flowers and seeds quite freely. However, when man intervenes such structures can be made to be very efficient means of propagation.

There are some disadvantages with the use of clonal varieties—for instance, it is not unusual for cultivars which do well in one climate to perform badly in another. This often happens with rose and Dahlia varieties raised in Europe and sent to other

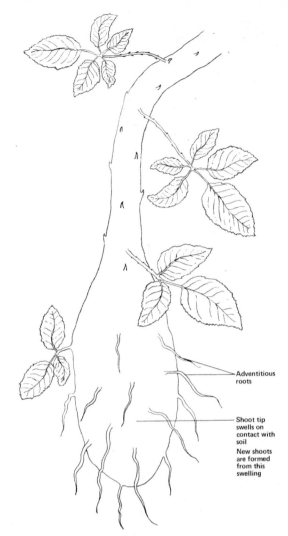

Adventitious roots

Shoot tip swells on contact with soil

New shoots are formed from this swelling

FIG 11.   BLACKBERRY STOLON

countries and *vice versa*. Similarly, individuals of clonal fruit cultivars produce different shaped fruit in different parts of the world which have different climates.

In addition, all the members of a clone will be equally susceptible to a specific strain of a disease. They will therefore be more severely affected if that strain occurs than would a seed-propagated variety with some variation regarding susceptibility.

8 Cuttings being prepared in a typical commercial potting shed.

9 General views of a modern nursery which is raising plants in containers. Note the extent of the operation and also the shade house at top centre.

6

Propagation by Cuttings

MATURE PLANT tissues are made up of many millions of individual cells. Each cell has a specific function and they are aggregated together to form distinct tissues or organs. The ability of a plant part to make a successful cutting depends upon the ability of cells within it to change their function, divide and produce new cells, which will in turn form new organs. Stem cuttings have to form new roots, root cuttings new shoots, and leaf cuttings, both new shoots and roots.

The processes involved in the formation of new tissues are complex and differ from species to species and on the plant part used. It is not possible or desirable to go into great detail here. However, it is worthwhile considering some of the underlying aspects of root initiation in shoot cuttings, as this is by far the most common form of vegetative reproduction.

Adventitious roots are formed actually within the stem from cells near to the vascular or conducting tissues. The process of their formation can be considered as falling into three stages:

1. The initiation of groups of meristematic cells. These are cells with the ability to divide and form new cells and tissues.
2. The development of these into root primordia (embryo roots).

3. The development and emergence of the new roots including the rupture of existing stem tissues and the formation of connections with the stem conducting tissues.

Slightly different tissues are involved in different plants, but in most instances the process starts slightly after the cutting is made and placed under favourable rooting conditions.

During rooting, callus, an irregular mass of cells develops on any cut surface. Contrary to common belief this callus formation is incidental to root formation and the adventitious roots simply push their way through it.

The Use of Hormone Preparations

All plant growth is controlled by growth substances. These fall into a number of distinct chemical groups, each of which has a specific function. They occur in all plants and their action is complex and still not fully understood. Some of these natural materials have been found to have the function of controlling and stimulating root formation. In addition, a number of synthetic substances have also been shown to be capable of stimulating and increasing root formation.

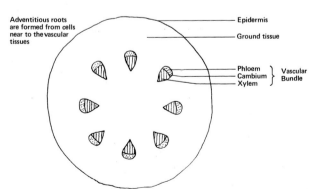

FIG 12a.   TRANSVERSE SECTION OF YOUNG HERBACEOUS STEM

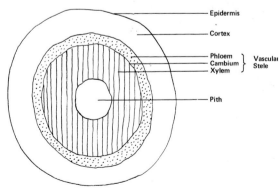

FIG 12b.   TRANSVERSE SECTION OF OLDER, MORE WOODY STEM

These discoveries were first made during the 1930s and it was not long before commercial preparations containing various combinations of root stimulating growth substances were made available to horticulturalists. It is now standard practice to dip the cut ends of cuttings in a hormone preparation before inserting them into a rooting medium. Some of the preparations currently available have fungicides as part of their formulation. This helps to combat fungi which might otherwise destroy a cutting before rooting could take place.

Naturally-occurring growth substances are synthesised primarily in apical buds and young leaves. It is therefore interesting to note that in a number of plants rooting will not take place when all the leaves are removed from the cuttings, even in the presence of added external hormone. Clearly, natural growth substances are more complex than synthetic ones.

Internal Factors Affecting the Rooting of Cuttings

It should always be borne in mind that a root system developed from a cutting differs in shape from that developed by a plant when it grows from a seed. It is in essence an artificial structure, although nonetheless efficient for that.

Plants are extremely variable in their ability to form such an adventitious root system. In some it is virtually impossible to induce them to do so, while in others even trimmings left on the ground will root freely. Most plants fall somewhere between these extremes and careful consideration of the factors influencing rooting and suitable action in the light of these, can result in a high level of success.

Broadly, cuttings should be taken from healthy, well-fed plants. Propagating material should never be taken from diseased plants and only from starved plants under circumstances of special need.

In plants that are easily rooted from cuttings the age of the stock plant makes little difference, but sometimes with more difficult subjects age can be an important factor. Cuttings taken from young plants still in the juvenile phase will root much more quickly and easily than those taken from older plants.

The type of material used to make cuttings varies from plant to plant. Some subjects are propagated from the young succulent growing tips, others use material which is several years old. These may bear leaves and be actively growing or may be dormant and have no leaves. Slightly different formulations of hormone powder are made to suit different types of cutting. These are usually referred to by number: 1, for softwood or herbaceous cuttings; 2, for semi-hard wood; and 3, for hardwood cuttings.

Marked differences in rooting ability occur between individuals of a species and consequently between clones. This probably explains why beginners are sometimes successful at rooting plants which "the book" says cannot be propagated from cuttings.

In some cases the part of the plant from which cuttings are made is important. Often cuttings from terminal upright shoots produce plants quite different in shape to plants made from lateral branches. In fact some horizontal forms of certain plants owe their existence as separate entities to the selection of suitable propagating material.

Quite often, lateral shoots root more easily and quickly than terminal ones, and generally stems with a large pith cavity in the middle make less successful cuttings than more solid thinner ones.

Many books stress that only non-flowering or vegetative shoots should be used for cuttings. In very many cases this is not critical although for many species, better regeneration takes place when cuttings are taken before or after flowering. However, for some plants it is difficult to find shoots which are not flowering. There are only a very few cases where flowering shoots will not form roots.

Fashions seem to change with regard to how cuttings should be prepared. Older books were very keen on making cuttings with a "heel", a piece of older wood torn off with the cutting. Then for many years gardeners have been advised to make a clean cut directly beneath a node. But today, with the advent of hormone rooting powders, even internodal cuttings, where the cut is made anywhere between nodes, are successful. So it does not seem very important

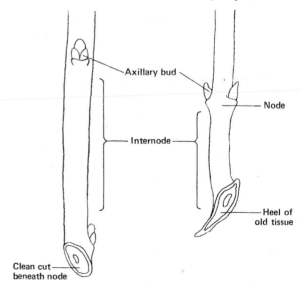

FIG 13a. NODAL CUTTING   FIG 13b. HEEL CUTTING

just where a cut is made for the vast majority of plants, and often the availability of propagating material becomes the over-riding consideration.

The time of the year when cuttings are made can have a considerable effect on the results, although it is probably the physiological state of the material at any given time and the method used that are important. Often plants grown in a glasshouse will produce far more successful cuttings than their counterparts growing outside. The time of the year and the type of cutting are of course linked. If dormant hardwood cuttings are to be struck directly in the open ground, they must be made and inserted somewhere between autumn and early spring. However, modern techniques such as mist propagation are in many cases altering old concepts.

A grower must always keep an open mind and a healthy suspicion of what he reads. Books such as this can only give a few clues; you should always be prepared to experiment in relation to your own set of circumstances.

Environmental Conditions

WATER SUPPLY

The maintenance of an adequate water supply to the tissues of a cutting while it is forming new roots presents the greatest problem to this form of propagation. The water supply to the leaves from the root has been cut off, but the leaves are still capable of losing water by transpiration, and unless special measures are taken the cutting will dry up and die before new roots are formed and the water supply re-established.

This is overcome in two major ways: first, the number of leaves on a cutting are reduced to a minimum, so that only the youngest are left. This ensures adequate production and translocation of food and growth substances, but reduces the leaf area from which water may be lost. The second measure involves maintaining a high humidity around the aerial parts of the cuttings. This has long been accomplished by the use of propagating frames, very similar to cold frames discussed earlier. These are situated in glasshouses and usually have their glass lightly painted with whitewash to reduce the light intensity within them. Often they have heating wires passing through sand which fills the bottom of them. These raise the temperature of the sand into which the cuttings are plunged, which both speeds up the process of rooting and also helps to maintain a high humidity in the air above the moist sand.

The advent of polythene sheet has enabled the construction of temporary propagating structures, both inside and outside glasshouses. For the amateur a polythene bag supported on sticks or wire hoops and tied over the top of a pot of cuttings often serves as an excellent propagating aid.

Traditionally, cuttings in propagating frames have been frequently sprayed with a syringe to obtain a high humidity.

A very important technique which has been developed since World War II is mist propagation, which was developed from devices designed to maintain a high humidity within a propagating structure. It has, however, several advantages over and above that of simply maintaining a high humidity around the leaves.

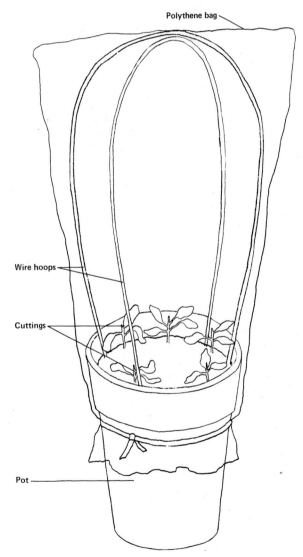

FIG 14.   MINIATURE PROPAGATING CASE

10 Commercial mist-propagation unit. The mist nozzles are spaced along the water supply pipe. Note the artificial leaf which controls the mist application in relation to climatic conditions.

The system consists basically of nozzles which produce a fine mist or fog linked to artificial leaves or sensors which switch the mist on every time their surface dries. This system maintains a film of water on the leaves which not only stops them from drying out, but also cools them and helps to maintain a more even temperature regime than that obtained in a closed frame. This cooling is so efficient that the cuttings do not have to be shaded and can thus make better use of the sunlight for producing foods and in turn grow more quickly. Larger cuttings may be used under mist, with the advantages of larger food reserves and a larger leaf area for food production, all of which result in better plants more quickly.

TEMPERATURE

Temperature is important in rooting cuttings. Most but not all plants root satisfactorily with an air temperature somewhere between 15-26°C (60-80°F). Too high air temperatures stimulate shoot growth before the roots are adequately formed, and for this reason the rooting medium is often maintained at between 26-38°C (80-100°F) by heating wires, which gives the roots a good start.

Rooting Media

Rooting media for cuttings have exactly the same job to perform as those used for seedlings, namely to anchor the cutting and resulting plant and to provide moisture and oxygen to the developing root system. The rooting medium should also be free of harmful organisms.

Different rooting media can produce different root systems, the moisture/oxygen availability determining the result. Generally speaking, where mist is employed straight sand is best as there is a regular supply of water, but under other circumstances the addition of peat helps to ensure a fairly uniform supply of moisture.

Main Types of Cuttings

Many different parts of plants can be used for cuttings, especially by a skilled propagator; however, the main types of cutting are as follows:
1. Stem cuttings
   a. Hardwood and semi-hardwood
   b. Softwood and herbaceous.
2. Leaf cuttings
3. Root cuttings.

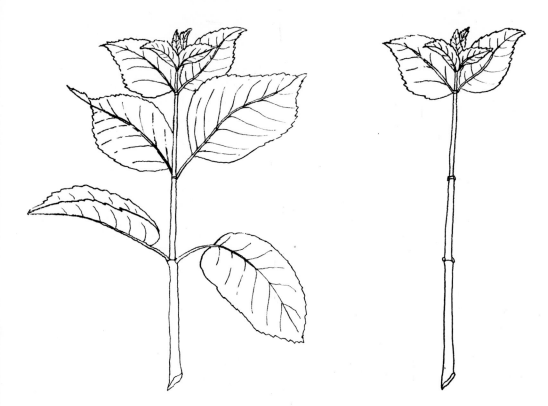

FIG. 15a. LEAFY SHOOT UNTRIMMED

FIG. 15b. CUTTING PREPARED FOR INSERTION INTO ROOTING MEDIUM

HARDWOOD AND SEMI-HARDWOOD CUTTINGS

Hardwood cuttings are the simplest and cheapest form of vegetative propagation. They are most often used for deciduous woody plants such as the rose understock species *Rosa multiflora*, privet, forsythia, willow and grape. Cuttings are prepared when the plants are dormant and have no leaves. Well ripened, one-year-old growth should be selected from vigorous healthy plants. Ideally, special plants should be reserved for the supply of cutting material. These then become stock plants and special attention can be paid to ensuring that the propagating material is free of disease and true to type.

Cuttings vary between four inches and six feet in length, depending on the plant and its intended use. The longest cuttings are used for the main stems of plants such as pendulous roses grown as standards. These are either budded or grafted to the stock plant after the cutting has become established. Such long cuttings need special precautions to prevent them from drying out before rooting can take place, but the more usual shorter lengths can be placed directly into the open ground any time from autumn to early

spring. Generally, autumn is best, as this gives plenty of time for root formation to become well established before aerial growth starts in the spring.

Preparation is not critical other than to ensure that each cutting has at least two buds or nodes from which shoots can develop. Where the plants are destined to make understocks on to which other varieties will be grafted, it is normal to remove all but the top two buds to prevent sucker formation later in the life of the plant.

Hardwood cuttings of evergreen conifers are generally more difficult to root than are deciduous plants, although certain *Chamaecyparis* and *Juniperus* species are not difficult. More difficult subjects are successfully rooted commercially by using hormone rooting powder and placing in a glasshouse, where high light intensities and humidities prevail. Bottom heat of 24-27°C (75-80°F) helps considerably in stimulating rooting. Once rooting has occurred, the plants are either planted out in lines in the field to grow into saleable plants or are put into containers. Autumn and winter are the normal times to take conifer cuttings. The home gardener can have a fair

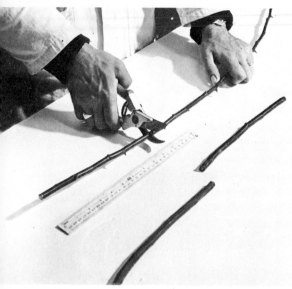

11 Dormant *Rosa multiflora* canes being cut to a standard 9-inch length.

12 All except the two top buds are removed and cuttings are pointed for easy insertion in soil.

13 The prepared cuttings being inserted in soil.

14 The hardwood cuttings lined out. In this case they have been inserted through a polythene mulch to stop weed growth, with sawdust between the lines.

measure of success with conifers by placing cuttings of the easier species directly into the open ground. He will need to be patient, however, as rooting can take over a year to occur and he will need to keep on top of weed control and watering to prevent the cuttings drying up and dying.

Quite a few plants are propagated from semi-hardwood cuttings. These are often broad-leaved evergreen plants such as camellia, pittosporum and euonymus. Such cuttings are taken mainly from new shoots resulting from a flush of growth. Soft imma-ture growth should, however, be avoided. Semi-hardwood cuttings are essentially of mature but leafy tissue.

Cuttings of semi-hardwood, softwood and her-baceous cuttings are prepared in a similar manner. They are usually made between three and six inches long and the larger lower leaves are removed in an attempt to reduce water loss. The basal cut is tra-ditionally made directly beneath a node, but this is not of vital importance, internodal cuttings perform-ing equally well. When propagating material is in short supply, more than one cutting may be made from a single shoot by leaving a node with its leaves at the top of successive segments, beneath the top cutting. Shoots will develop from the dormant axillary buds of the top node.

15 Stock plant of Fuchsia with a good supply of suitable softwood cuttings.

16 Cuttings being removed with scissors. The cut stump of the stock plant should be trimmed back close to the axil immediately below the cut. Two new shoots will replace the one removed.

17 All but the youngest leaves are removed to reduce water loss from the unrooted cuttings.

18 *Right:* The lower end of the cutting is dipped into hor-mone rooting powder and any surplus knocked off.

SOFTWOOD AND HERBACEOUS CUTTINGS

The distinction between these cuttings is largely one of definition or origin. Strictly, softwood cuttings are prepared from the soft spring growth of deciduous and evergreen shrubs and trees, many of which may also be propagated by hardwood cuttings. Herbaceous cuttings are made from the succulent growth of plants such as geraniums (pelargoniums), coleus and carnations. In addition, plants such as chrysanthemums and Dahlias produce herbaceous shoots from their stools and tubers and these are used for cuttings.

LEAF CUTTINGS

Here either the leaf blade, or the leaf blade together with the petiole (or leaf stalk), is used. For plants like African violet *(Saintpaulia sp.)*, a leaf and stalk is removed from the plant and the stalk pressed into a rooting medium up to the leaf blade. Roots form at the base of the stalk and new leaves are formed directly above them.

FIG. 16. LEAF CUTTING OF BEGONIA REX

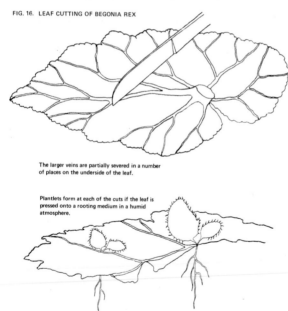

The larger veins are partially severed in a number of places on the underside of the leaf.

Plantlets form at each of the cuts if the leaf is pressed onto a rooting medium in a humid atmosphere.

For plants such as *Begonia rex* the large veins on the underside of mature leaves are partially (but not completely) severed. The cuts can be treated with hormone rooting powder, using a small brush. The leaves are then laid flat on a rooting mixture and pinned down with small wire pegs, or more usually, weighted down with small pebbles. Most of the leaf is, however, exposed to the light. After several weeks, under humid conditions, new plants will form at the points where each vein was cut. The old parent leaf blade eventually disintegrates.

ROOT CUTTINGS

These are not commonly used, although there are a number of quite well-known plants which may be propagated in this way—for instance, lilac, wisteria, phlox and plumbago. It is important, however, to distinguish between a true root cutting and an underground stem, which can often look very much like a root.

Making root cuttings of plants which can be propagated in this way is simple, but often the work of obtaining suitable material can be difficult. Roots are simply cut into segments and either placed horizontally just beneath the soil surface or with the cut end which was nearest the crown of the parent plant uppermost. After a few weeks new roots and shoots are formed.

Hardening-Off

It is important, once cuttings raised under mist or high humidity have rooted, to harden them off or to wean them in the same way as seedlings. When mist propagation is employed the intervals between mist applications can be increased until it is left off altogether, or alternatively, if the cuttings have been placed in pots or boxes rather than into a propagating bench, the plants can be moved gradually farther away from the misting units and watered only as required.

Where propagating frames are used, humidity can be progressively reduced by increasing the ventilation.

Frequently, rooted plants taken from a propagating bench are potted-up and grown in a fairly warm humid atmosphere until established. They are then moved either to a cooler glasshouse, cold frame or lathhouse, until planted outside or sold.

7

Other Simple Forms of Vegetative Propagation

CUTTINGS are the single most important form of vegetative reproduction employed by man. There are, however, an enormous number of techniques and variations of vegetative reproduction which may be employed. Some of these are simple and well within the reach of the ordinary home gardener, others are of interest only to specialist horticulturalists, and still more are little more than research techniques.

In a book of this size it is only possible to look at some aspects of the subject and for this reason the simplest forms of vegetative propagation other than cuttings will be considered. Many of these make use of natural plant modifications.

Runners

A runner is a specialised stem which develops from the axil of a leaf at the crown of a plant, grows horizontally along the surface of the ground and forms new plants at some of the nodes along its

length. The strawberry is the best known example of this form of reproduction. Rooted daughter plants may be lifted from where they have rooted and re-planted or they may be encouraged to root directly into a sunken flower pot. This method lessens the shock of transplanting, although it would be impracticable on a commercial scale.

Stolons

This term covers a range of stems which have the ability to form roots when coming into contact with the soil. Tip layers of blackberry present a good example which may be treated in the same way as strawberry runners.

Offsets, Suckers and Division

In practice the terms "sucker" and "offset" are used loosely to describe any shoot that is produced from a plant part below ground. Offsets are usually

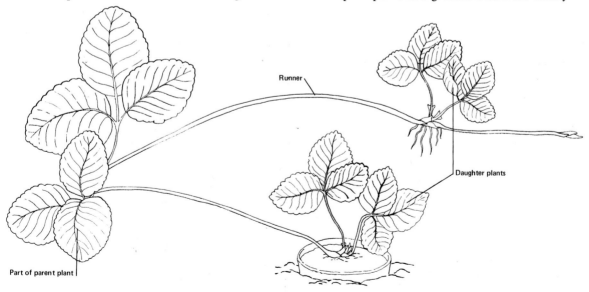

FIG 17.   STRAWBERRY RUNNERS MAY BE ENCOURAGED TO FORM
DAUGHTER PLANTS DIRECTLY INTO POTS

41

taken to mean shoots formed close to an existing main stem. Some authors, however, have given the terms quite specific meanings.

Many herbaceous perennial plants produce lateral shoots, which develop from the base of the main stem. Each year new lateral shoots are formed from the base of the previous year's shoots, and after a while a clump, crown or stool develops.

Offsets may be removed singly from the parent plant with a sharp knife or spade and treated like a rooted cutting. These are sometimes referred to as "Irishman's cuttings" as they are already rooted. Alternatively, where a large stool or clump has been formed, the whole plant may be lifted and divided up into smaller clumps for replanting elsewhere.

Most bulbs and corms produce offset bulblets and cormels. Each of these, if separated, will grow independently and in time produce a mature flowering plant.

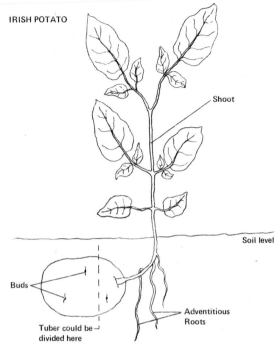

IRISH POTATO

Shoot

Soil level

Buds

Adventitious Roots

Tuber could be divided here

FIG 18. LONGITUDINAL SECTION OF GLADIOLUS CORM SHOWING CORMLET FORMATION

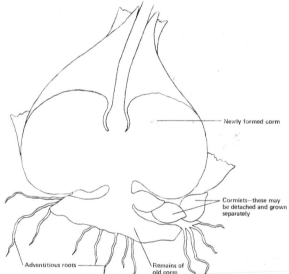

Newly formed corm

Cormlets—these may be detached and grown separately

Adventitious roots

Remains of old corm

FIG. 19a, 19b. DIVISION OF UNDERGROUND STEMS

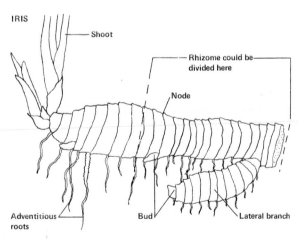

IRIS

Shoot

Rhizome could be divided here

Node

Adventitious roots

Bud

Lateral branch

Underground Stems

Underground stems include tubers and rhizomes. A tuber is strictly an underground stolon which has become swollen and contains food reserves, while a rhizome is a specialised stem which grows horizontally just beneath the surface of the soil. The Irish potato (*Solanum tuberosum*) is the best example of a stem tuber, and lily-of-the-valley and Iris present good examples of rhizomes. Both structures have all the parts of an aerial stem, modified to suit the radically different subterranean environment.

Most tubers and rhizomes may be increased simply by cutting them up into segments, provided that each segment has buds and is large enough to contain adequate food reserves. Aerial shoots are produced from these buds and adventitious roots form mainly on the bases of these aerial shoots.

Tuberous Roots

In some herbaceous perennial plants, all the aerial growth dies down at the end of the season and no living part remains above ground. In such plants the roots may become swollen and contain food storage materials. Swollen roots may be distinguished from underground stems as they have no nodes or internodes and buds are present only at the top at the crown or point where last year's stem joins the root. Adventitious roots are formed directly on their lower ends. Sweet potato, tuberous begonia and Dahlia are good examples of plants with tuberous roots. Sweet potato and Dahlia tubers differ from those of tuberous begonias, however, in that they are swollen lateral roots, which function for only one or two seasons, while the begonia tuber is formed from the swollen tap root and is perennial, growing larger each season.

As with underground stems, plants with root tubers may be increased simply by division, provided each piece of storage tissue has a bud attached to it. Often it is a good idea to wait until these buds have started to grow a little before making the division. It is then much easier to see what you are doing, although extra care must be taken not to damage the buds at that stage.

Both underground stems and tuberous roots are commonly lifted and provided with bottom heat so that adventitious roots are produced and aerial growth stimulated. The shoots produced are then

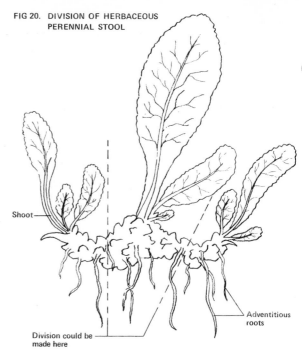

FIG 20. DIVISION OF HERBACEOUS PERENNIAL STOOL

Shoot

Adventitious roots

Division could be made here

removed and prepared as herbaceous cuttings. In plants such as Dahlias the crown of the plant, with the buds, is left well clear of the peat and sand rooting mixture, while in sweet potato the whole tuber is covered. In this instance, when shoots are produced, they form roots at their bases and are removed already rooted.

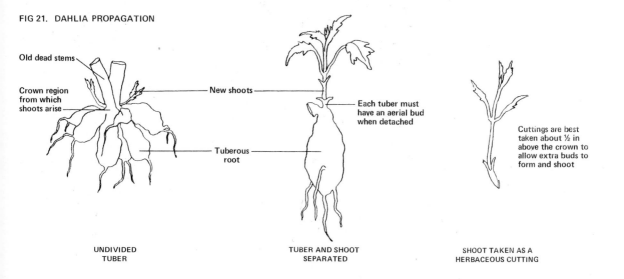

FIG 21. DAHLIA PROPAGATION

Old dead stems

Crown region from which shoots arise

New shoots

Each tuber must have an aerial bud when detached

Tuberous root

Cuttings are best taken about ½ in above the crown to allow extra buds to form and shoot

UNDIVIDED TUBER

TUBER AND SHOOT SEPARATED

SHOOT TAKEN AS A HERBACEOUS CUTTING

Layering

Layering is the development of roots on a stem while it is still attached to the parent plant. Runners and stolons, which we have already discussed are natural forms of layering.

Quite a number of plants can be induced to produce "layers". Some simply need to have an aerial shoot pulled down and a short section of it pegged down beneath the surface of the soil. Others give best results if the section pegged down is half-severed and is treated with hormone rooting powder.

The advantage of layering over making cuttings lies in the fact that the layer is supported by the parent plant until roots have become established and it can support itself. Plants difficult or slow to form roots as cuttings often can be satisfactorily propagated in this way.

A disadvantage of the method is that generally many more cuttings than layers can be obtained from a parent plant; the system is hardly used commercially.

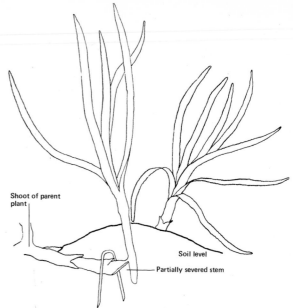

Shoot of parent plant

Soil level

Partially severed stem

FIG 22. CARNATION LAYER

19 Plants in plastic bag containers standing on gravel for growing-on. The slatted fence in the background acts as a windbreak. The neatness and cleanliness of the area enable good disease control and easy access to the plants.

20 Plants being fed with liquid fertiliser through a pump and hose system on a weekly routine.

21 Container-grown plants are watered automatically by a permanent sprinkler system laid out as a grid. This eliminates an enormous amount of tedious and expensive hand labour.

22 Typical range of glasshouses in a modern nursery. Note the sealed access way and general cleanliness.

23 House plants being grown-on under glass. Note again the absence of anything superfluous and the complete cleanliness.

PART III

8

Weed and Disease Control

THE SUCCESS or failure of propagation, whether sexual or vegetative, is often determined by the control of harmful organisms.

Weeds, especially in outdoor sowings, can easily outgrow and swamp small seedlings. Nematodes, insects and pathogenic fungi and bacteria readily attack seedlings and cuttings. In addition, vermin can cause problems in glasshouses and frames, unless controlled.

Hygiene is the key to all disease control, and I cannot emphasise this point too strongly. If disease organisms are denied places in which they can lurk, most of our problems will be avoided.

In practice this involves little more than applying to our gardens and nurseries the standards of cleanliness and tidyness which we have come to accept as normal in our homes.

It is important to:
1. Keep propagating areas free of "junk" (i.e. anything not directly concerned with propagating).
2. Not to throw rubbish and plant parts on the floor or under the bench.
3. To burn all diseased and woody plant material as soon as possible and to compost the rest.
4. To use only clean tools, containers and labels.
5. To wash propagating structures out regularly and to clean containers before storage.

If these elementary points are observed the garden or nursery becomes tidier, more efficient, more pleasant to work in and above all more likely to avoid serious disease problems. Time spent regularly on cleaning is a far better investment than many dollars spent on expensive and frequently misused chemical sprays.

Soil presents the greatest single danger to successful propagation, both under glass and in the open. It contains many beneficial organisms, both large and microscopic, but, as taken from the field, it also contains weed seeds and some harmful organisms. Many diseases are capable of forming special resting structures which can exist in soil for many years until suitable conditions occur, which enable them to attack a crop. "Damping-off", the condition where young seedlings suddenly collapse, is caused by soil fungi such as Pythium. This is one of the reasons why soilless seed and potting composts are gaining in popularity.

Where soil is used in seed and potting composts it is normal to treat it either with heat, usually in the form of steam, or with chemicals. Similarly, the soil of glasshouse beds and special areas in the open are often fumigated to eliminate weeds, nematodes and harmful fungi and bacteria.

Care should be taken when treating soil, as excessive use of either heat or chemicals can cause troubles. Too much heat can lead to excessive breakdown of the organic fraction of soil, a build-up of soluble salts and the production of toxic chemicals. Similarly, the use of chemicals needs to be properly understood, or damage can occur to neighbouring plants and residues left in the soil can make soil unusable until they have been removed. In addition, if soil is sterilised the whole flora and fauna of the soil are destroyed, which means that the first organisms to colonise it after treatment can develop without their normal competitors and population explosions occur. If the first organisms happen to be disease ones the results are disastrous. Brief details of soil treatments are given in the appendix.

After the soil, the propagating material itself is the next most important source of disease. Seeds are capable of carrying fungal and bacterial diseases, both on their surfaces and within their tissues. For this reason hot water treatments where seeds are suspended in water at temperatures between 49-57°C (120-135°F) for periods up to half an hour, are sometimes employed for certain types of seed. In other cases, seeds may be treated with a chemical which has the effect of disinfecting the seed and also of protecting the seedling immediately after germination.

With vegetative propagation it is vitally important to propagate only from parent plants known to be free of disease. Many virus diseases and some fungal and bacterial diseases are systemic and show symptoms only under certain conditions. For big commercial undertakings it is often necessary and worthwhile to carry out specific tests to determine the health status of propagating material. Similarly, when propagating material is imported into a country, quarantine requirements have to be met. These are formulated in an attempt to prevent the importation of diseases new to that country. Any disease invariably causes reductions in yield and general performance, although the extent of this reduction is often not fully appreciated.

The amateur gardener can often determine the status of his parent material by careful observation throughout several seasons. Any plants showing disease symptoms should either be marked and avoided when propagating, or destroyed.

APPENDIX A

Plant Lists

THE FOLLOWING lists are intended as a guide to the normal propagation methods employed by nurserymen in other countries. The limitations of these must, however, be appreciated if they are not to be misleading.

Space allows only genera to be listed. Often a single genus will embrace a number of different species and perhaps several hundred cultivars. For this reason, it may be normal to grow some species of a genus from seed, while hybrid forms of the same genus will have to be maintained as clones by vegetative means. For example acers may be readily grown from seed, but the choicest forms are normally budded or grafted on to seedlings of *Acer palmatum*. In such cases both forms of propagation are indicated. In other cases two methods may be equally practicable and here also a double entry is made. A grower may want all his plants to be identical, in which case he will choose the vegetative methods, while if he wants variation and the chance to find new forms, he must raise his plants from seed.

Because only one or perhaps two methods are listed for any entry, this does not mean that other methods are not possible. It is surprising what can be done with plant material if one has time and patience. The list will simply tell you the method to try first.

Common names have not been listed, except for fruit, as they usually apply to individual species rather than to genera. In addition they often have only very local currency and may thus be misleading.

I acknowledge the help and guidance received from Mr George Rainey, President of the New Zealand Nurserymen's Association, in compiling the list of trees and shrubs, and Mr David Goudie of Zealandia Nurseries, Auckland, for supplying me with details of the methods generally used for propagating house plants.

Key to Abbreviations

S	=	seed
C	=	stem cutting
RC	=	root cutting
LC	=	leaf cutting
B	=	budding
Gr	=	grafting
D	=	division
HC	=	hardwood cutting
L	=	layer
HW	=	hot water soak for hard-coated seeds. Chipping may be employed for large seed.

List of Common Tree and Shrub Genera

Abelia	HC	Aucuba	S/C
Abutilon	S/C	Azalea	S/C
Acacia	S (HW)	Banksia	S
Acer	S/Gr	Beloperone	C
Agathis	S	Berberis	S/C
Agave	S/D	Betula	S/Gr
Agonis	S	Boronia	S/C
Ailanthus	S	Buddleia	C
Albizia	S	Callistemon	S/C
Araucaria	S	Calluna	C
Arbutus	S	Camellias	C
Asclepias	S/C	Cassia	S(HW)/C

49

Casuarina	S	Hoheria	S/C
Catalpa	S/RC	Hydrangea	C
Ceanothus	C	Hypericum	C
Cedrus	S/Gr	Idesia	S
Cercis	S	Ilex	S/C
Cestrum	S/C	Jacaranda	S
Ceratostigma	S/C	Jacobinia	C
Chaenomeles	S/C	Juniperus	C
Chamaecyparis	C	Kalmia	S
Choisya	C	Laburnum	S/Gr
Clerodendrum	S/C/RC	Lantana	C
Clethra	S/C	Leonotis	C
Clianthus	S	Leptospermum	S/C
Coleonema	C	Leucadendron	S/C
Coprosma	S/C	Leucospermum	S/C
Cordyline	S	Ligustrum	S/C
Cotinus	S/L	Liquidamber	S/B
Cotoneaster	S/C	Liriodendron	S/B
Cryptomeria	S/C	Luculia	S/C
Cuphea	C	Magnolia	S/C/L
Cupressus	S/C	Mahonia	S
Cyathea	(Spores)	Malus	B
Cyperus	D	Melaleuca	S/C
Cytisus	S/C	Metrosideros	S/C
Dacrydium	S	Michelia	S/C/L
Daphne	C	Monstera	S/C/D
Datura	C	Musa	S/D
Deutzia	C	Nerium	S/C
Diosma	C	Nothofagus	S
Dryandra	S	Olearia	S/C
Embothrium	S	Paulownia	S/RC
Epacris	S/C	Pernettya	C
Erica	S/C	Phebalium	C
Eriostemon	C	Philadelphus	C
Escallonia	C	Philodendron	S/C
Eucalyptus	S	Phoenix	S
Eugenia	S	Phormium	S/D
Euonymus	C	Photinia	S/C
Euphorbia	S/C	Phylica	S
Euryops	C	Picea	S/Gr
Exochorda	C	Pieris	S/C
Fagus	S/Gr	Pinus	S/Gr
Ficus	S/C	Pittosporum	S/C
Forsythia	C	Plumbago	C
Fraxinus	S/B	Podalyria	S/L
Fremontia	S	Podocarpus	S/C
Fuchsia	C	Polygala	C
Garrya	C	Populus	C
Ginkgo	S	Prostranthera	C
Grevillea	S/C	Protea	S/C
Griselinia	C	Prunus	S/C/B/Gr
Hakea	S	Pseudopanax	S
Hebe	C	Pseudotsuga	S
Hibiscus	C	Punica	S/C

Pyracantha	S/C	Swainsona	S/C
Quercus	S	Syringa	C/B
Romneya	S/RC	Tamarix	HC
Rhododendron	C/L	Taxodium	S
Ribes	C	Taxus	C
Ricinus	S	Telopea	S
Rondeletia	C	Tetrapanax	RC
Rosa	S/C/B	Teucrium	C
Rosmarinus	C	Thryptomene	C
Russelia	C	Thuja	C
Salix	C	Tibouchina	C
Schinus	S	Tweedia	S
Sequoia	S	Ulmus	C/B
Sesbania	S	Viburnum	C
Sophora	S	Virgilia	S
Sorbus	S/B	Vitex	S/C
Spirea	C	Weigela	C
Stenocarpus	S	Zauschneria	C
Strelitzia	S/D		

List of Common Perennial Climbing Plants

Aristolochia	S	Manettia	C
Bignonia	S/C	Metosideros	S/C
Bomaria	S/D	Pandorea	C
Bougainvillea	C	Parthenocissus	C
Campsis	S/C/RC	Passiflora	S/C
Celastrus	C	Phaedranthus	C
Clematis	C/S	Phaseolus	S(HW)
Convolvulus	C	Pyrostegia	C
Ficus	C	Solanum	C
Gelsemium	C	Sollya	S/C
Hardenbergia	C/S	Stauntonia	S/C
Hedera	C	Stephanotis	C
Jasminium	C	Thunbergia	S/C
Lapageria	S	Tecoma	C
Lathyrus	S(HW)	Trachelospermum	C
Lonicera	C/Gr	Wisteria	C
Mandevilla	S		

List of Main Fruit Plants

Almond	B	(Meyer Lemon	C)
Apple	B	Cranberry	C
Apricot	B	Currants	HC
Boysenberry	D/L	Feijoa	S/C
Chinese		Fig	HC
Gooseberry	C	Gooseberry	HC
Citrus	B	Grape	C/Gr

Guava	S	Plum	B
Loganberry	D/L	Quince	B
Macadamia	S	Raspberry	D
Nectarine	B	Rhubarb	S/D
Passion Fruit	S/C/Gr	Strawberry	Runner
Paw Paw	S	Tree Tomato	S/C
Peach	B	Walnut	S/B/L
Pear	B		

(It should be noted that where fruit plants are budded or grafted, the understock often has a considerable effect on the performance of the resulting plant. For this reason it is important to obtain the appropriate understock.)

List of Main House Plants

Abutilon	C	Gardenia	C
Acalypha	C	Gloxinia	
Achimenes	C/D	(Sinningia)	S
Adiantum	S	Hedera	C
Aeschynanthus	C	Hoya	C
Anthurium	S/D	Isolepis	
Aphelandra	C	(Scirpus)	D
Araucaria	S	Kalanchoe	C
Ardisia	C	Kentia	S
Asparagus	S	Maranta	D
Asplenium	S	Monstera	S/C/D
Aucuba	C	Musa	S/D
Begonia	S/C/LC	Nephrolepis	D
Brassaia	S	Panicum	C
Calathea	D	Pelargonium	C
Calceolaria	S	Peperomia	C
Chamaedorea	S	Philodendron	S/C
Chlorophytum	D	Phyllitis	S
Chrysanthemum	C	Pilea	C
Cineraria	S	Pisonia	C
Cissus	C	Poinsettia	
Clerodendrum	C	(Euphorbia)	C
Coleus	S/C	Primula	S
Crossandra	C	Pteris	S
Columnea	C	Ravenala	S
Cordyline	S/C	Rhoeo	S/C
Croton	C	Saintpaulia	LC
Ctenanthe	D	Sanseveria	LC/D
Cyclamen	S	Schlumbergera	LC
Cyperus	D/S	Selaginella	C
Dieffenbachia	C	Sonerila	C
Dizygotheca	S	Stromanthe	D
Dracaena	C	Strobilanthes	C
Episcia	C	Tolmiea	Plantlet forms
x Fatshedera	C		on old leaf stalk
Fatsia	S	Tradescantia	C
Ficus	S/C/L	Vriesia	S
Fittonia	C	Zebrina	C
Fuchsia	C		

Common Annual Plants Raised from Seed

Ageratum
Agrostemma
Alyssum (Lobularia)
Amaranthus
Anagallis
Anchusa
Antirrhinum
Artotis
Aster (Callistephus)
Begonia
Calendula
Capsicum
Celosia
Centaurea
Cerinthe
Cheiranthus
Chrysanthemum
Clarkia
Cleome
Convolvulus
Cosmos
Dahlia
Delphinium (Larkspur)
Dianthus
Dimorphotheca
Emilea
Eschscholtzia
Exacum
Godetia
Gomphrena
Gypsophila
Helianthus (Sunflower)
Helichrysum
Heliotropium
Iberis (Candytuft)
Impatiens

Ipomea
Kochia
Lathyrus (Sweet Pea)
Layia
Linaria
Linum
Lobelia
Malcomia
Matthiola
Mesembryanthemum
Mimulus
Molucella
Myosotis
Nemesia
Nigella
Papaver
Penstemon
Petunia
Phacelia
Phlox
Portulaca
Rudbeckia
Salpiglossis
Salvia
Sanvitalia
Scabiosa
Schizanthus
Tagetes
Tithonia
Tropaeolum
Ursinia
Verbena
Vinca
Viola (Pansy)
Zinnia

Common Herbaceous Perennial Plants Which May Readily Be Grown from Seed

Acanthus
Achillea
Aquilegia
Anemone
Aster (Michaelmas Daisy)
Auricula
Campanula
Chrysanthemum
Canna
Coreopsis

Delphinium
Erigeron
Eryngium
Freesia
Gesneria
Gaillardia
Gerbera (Seed short lived)
Geum
Helenium
Hollyhock (Althaea)

Inula
Kniphofia
Liatris
Lupinus
Lychnis
Lythrum
Matricaria
Meconopsis

Nepeta
Primula
Pyrethrum
Ranunculus
Solidago
Trollius
Verbascum

Common Rock Plants Easily Raised from Seed

Allium
Alyssum
Anagallis
Androsace
Anthemis
Arabis
Arenaria
Armeria
Aubrieta
Campanula
Dianthus
Draba
Erinus
Genista
Gentiana
Geranium

Geum
Helianthemum
Iris
Leontopodium (Edelweiss)
Linum
Lithospermum
Nepeta
Oenothera
Potentilla
Primula
Saxifraga
Sedum
Sempervivum
Silene
Thymus
Viola

APPENDIX B

Seed and Potting Compost Formulae

John Innes Seed Compost (Proportions by volume, in metric units, with Imperial equivalents)

2 parts loam	*2 parts loam*
1 part peat	*1 part peat*
1 part sand	*1 part sand*
2 lb superphosphate	*1200 gm superphosphate*
1 lb ground limestone	*600 gm ground limestone*
per cubic yard	per cubic metre

John Innes Potting Compost

7 parts loam	*7 parts loam*
3 parts peat	*3 parts peat*
2 parts sand	*2 parts sand*
1 lb ground limestone	*600 gm ground limestone*
5 lb John Innes Base	*3,000 gm John Innes Base*
which consists of:	which consists of:

2 parts by weight—hoof and horn
2 parts by weight—superphosphate
1 part by weight—potassium sulphate
   per cubic yard   per cubic metre

Loam is taken to mean (in the strict original sense) turves which have been stacked for a period varying between 9 and 12 months and allowed to decompose. However, good fibrous topsoil is most often used. This should be steam-sterilised and sieved before mixing. Sand particles should be between 1/8 and 1/16 in or 1.5-3 mm in size.

University of California Mixes
(A number of variations have been formulated for different uses. Details of these may be found in *The U.C. System for Producing Healthy Container-Grown Plants* (see book list). However, the formula listed below has been found to be satisfactory for a wide range of seedlings and cuttings.)

50% sand (No.2 grade)	*50% sand (No.2 grade)*
50% peat	*50% peat*
fertiliser per cubic yard	fertiliser per cubic metre
4 oz potassium nitrate	*150 gm potassium nitrate*
4 oz potassium sulphate	*150 gm potassium sulphate*
40 oz superphosphate	*1,500 gm superphosphate*
120 oz dolomite lime	*4,500 gm dolomite lime*
40 oz agricultural lime	*1,500 gm agricultural lime*
40 oz Uramite	*1,500 gm Uramite*

APPENDIX C

Chemicals Used for Treating Soil and Potting Mixtures

Formaldehyde (Formalin 40%)
Kills fungi and some weed seeds. Applied as a drench. 1 part of 40% formalin in 50 parts water at a rate of between 5 and 10 gallons per square yard or between 25 and 50 litres per square metre. Soil should be covered for 24 hours after application. Soil must be aerated for at least 2 weeks after treatment.

Metham (Metam, Vapam, Winpam; Sodium-N-methyldithiocarbamic acid)
Liquid. Use as per label instructions. It decomposes in soil to give off gas composed mainly of methyl isothiocyanate.

Dazomet (DMTT, Basamid, Basamid Granular, Mylone, Boots Soil Steriliser, IWD Soil Disinfectant; Tetrahydro-3, 5-dimethyl-2H-1, 3, 5-thiadiazine-2-thione)
Powder or granules. Worked into soil mechanically. Breaks down to give off gas, again mainly composed of methyl isothiocyanate.
More pleasant and easier to use than liquid formulations. Follow the instructions on the packet carefully.

Chloropicrin (Tear gas; Trichloronitromethane)

Methyl bromide
Both of the above materials can be very useful soil fumigants, but both are dangerous to use and can harm both plants and humans. They should be handled only by professional horticulturalists or trained operators.
There are some preparations which consist of mixtures of chloropicrin and methyl bromide.

ADDITIONAL NOTE
Benomyl (Benlate) is a very effective fungicide for the control of *Botrytis cinerea* or Grey mould. It is therefore very useful in propagating frames, where this fungus can cause great damage.

APPENDIX D

Temperature Conversion Chart

Look at the central column and read off either the Celsius or Fahrenheit equivalent in the appropriate column.

°F		°C			
32	0	−17.8	91.4	33	0.6
33.8	1	−17.2	93.2	34	1.1
35.6	2	−16.7	95	35	1.7
37.4	3	−16.1	96.8	36	2.2
39.2	4	−15.6	98.6	37	2.8
41	5	−15	100.4	38	3.3
42.8	6	−14.4	102.2	39	3.9
44.6	7	−13.9	104	40	4.4
46.4	8	−13.3	105.8	41	5
48.2	9	−12.8	107.6	42	5.6
50	10	−12.2	109.4	43	6.1
51.8	11	−11.7	111.2	44	6.7
53.6	12	−11.1	113	45	7.2
55.4	13	−10.6	114.8	46	7.8
57.2	14	−10	116.6	47	8.3
59	15	−9.4	118.4	48	8.9
60.8	16	−8.9	120.2	49	9.4
62.6	17	−8.3	122	50	10
64.4	18	−7.8	123.8	51	10.6
66.2	19	−7.2	125.6	52	11.1
68	20	−6.7	127.4	53	11.7
69.8	21	−6.1	129.2	54	12.2
71.6	22	−5.6	131	55	12.8
73.4	23	−5	132.8	56	13.3
75.2	24	−4.4	134.6	57	13.9
77	25	−3.9	136.4	58	14.4
78.8	26	−3.3	138.2	59	15
80.6	27	−2.8	140	60	15.6
82.4	28	−2.2	141.8	61	16.1
84.2	29	−1.7	143.6	62	16.7
86	30	−1.1	145.4	63	17.2
87.8	31	−0.6	147.2	64	17.8
89.6	32	0	149	65	18.3

150.8	66	18.9	210.2	199	37.2
152.6	67	19.4	212	100	37.8
154.4	68	20	213.8	101	38.3
156.2	69	20.6	215.6	102	38.9
158	70	21.1	217.4	103	39.4
159.8	71	21.7	219.2	104	40
161.6	72	22.2	221	105	40.6
163.4	73	22.8	222.8	106	41.1
165.2	74	23.3	224.6	107	41.7
167	75	23.9	226.4	108	42.2
168.8	76	24.4	228.2	109	42.8
170.6	77	25	230	110	43.3
172.4	78	25.6	231.8	111	43.9
174.2	79	26.7	233.6	112	44.4
176	80	20.7	235.4	113	45
177.8	81	27.2	237.2	114	45.6
179.6	82	27.8	239	115	46.1
181.4	83	28.3	240.8	116	46.7
183.2	84	28.9	242.6	117	47.2
185	85	29.4	244.4	118	47.8
186.8	86	30	246.2	119	48.3
188.6	87	30.6	248	120	48.9
190.4	88	31.1	249.8	121	49.4
192.2	89	31.7	251.6	122	50
194	90	32.2	253.4	123	50.6
195.8	91	32.8	255.2	124	51.1
197.6	92	33.3	257	125	51.7
199.4	93	33.9	258.8	126	52.2
201.2	94	34.4	260.6	127	52.8
203	95	35	262.4	128	53.3
204.8	96	35.6	264.2	129	53.9
206.6	97	36.1	266	130	54.4
208.4	98	36.7	267.8	131	55
					55.6

Books for Supplementary Reading

Hartmann, Hudson T. & Dale, E. Kester. 1968. *Plant Propagation, Principles and Practices*, 2nd Edition; Prentice Hall, New Jersey, USA.

Harrison, Richmond E. 1963. *Handbook of Trees & Shrubs for the Southern Hemisphere*, 3rd Edition; Harrison, Palmerston North, New Zealand.

Garner, R.J. 1967. *The Grafters Handbook*, 3rd Edition; Faber & Faber, London, UK.

Baker, Kenneth F. (editor). 1957. *The U.C. System for Producing Healthy Container-Grown Plants*, Manual 23; California Agricultural Experimental Station Extension Service, University of California, USA.

Menage, R.H. 1964. *Introduction to Greenhouse Gardening*, Phoenix House, London, UK.

INDEX

MORE GARDEN BOOKS FROM DRAKE PUBLISHERS INC

DICTIONARY OF ANNUAL PLANTS	$5.95
BEGINNER'S GUIDE TO HYDROPONICS	$5.95
HERBS	$5.95
THE PERFECT VEGETABLE AND HERB GARDEN	$7.95
DRIED FLOWERS FOR DECORATION	$5.95
BONSAI	$5.95
MUSHROOM GROWING	$4.94
IKEBANA	$6.95
AN ALL-YEAR-ROUND GARDEN	$5.95
BASIC GARDENING	$4.95
BEGINNER'S GUIDE TO ROSE GROWING	$6.95
HOW TO GROW POTTED PLANTS	$4.95
ORCHIDS	$7.95
RHODODENDRONS	$7.95
POPULAR FLOWERING PLANTS	$5.95
POPULAR FLOWERING SHRUBS	$5.95